Race Car Vehicle Dynamics Problems, Answers and Experiments

Douglas L. Milliken
Edward M. Kasprzak
L. Daniel Metz
William F. Milliken

Warrendale, Pa.

SAE Permissions
400 Commonwealth Drive
Warrendale, PA 15096-0001 USA
E-mail: permissions@sae.org
Tel: 724-772-4028
Fax: 724-772-4891

SAE
400 Commonwealth Drive
Warrendale, PA 15096-0001 USA
www.sae.org/bookstore
E-mail: CustomerService@sae.org
Tel: 877-606-7323 (inside USA and Canada)
724-776-4970 (outside USA)
Fax: 724-776-1615

ISBN-10 0-7680-1127-2
ISBN-13 978-0-7680-1127-2

Library of Congress Control Number: 2002111179

SAE Order No. R-280

Printed in the United States of America.

Table of Contents

List of Figures

Introduction

In 1997, the SAE Publications Department realized that *Race Car Vehicle Dynamics* (RCVD) was being used as a university textbook. Based on comments from faculty, SAE requested that we prepare a supplementary book of problems. The *RCVD Workbook* was first published in 1998 and is being used in a number of vehicle dynamics courses. We also prepared a solutions manual which was supplied in very limited quantities to university faculty only, in photocopy form.

RCVD is still on the SAE best-seller list and we realized that many users outside the university environment were interested in both the problems and the worked solutions. Also, since the *RCVD Workbook* has been around for a few years, faculty have now prepared their own exam questions. This book combines new material with most of the original *RCVD Workbook* and solutions manual.

Updates and improvements in the current volume have been largely done by Douglas Milliken and Edward Kasprzak, while material by the senior authors Dr. L. Daniel Metz and William Milliken stands essentially as it was first written. There are many new problems and solutions which came primarily from the Road Vehicle Dynamics elective course given by Dr. William Rae and, more recently, by Edward Kasprzak at the University at Buffalo (UB). As with both RCVD and *Chassis Design: Principles and Analysis*, illustrations were prepared by Robb Ramsey. Typesetting was done by Mr. Kasprzak using the TeX software package.

The *Race Car Vehicle Dynamics Program Suite* on the accompanying CD has proven to be a valuable learning tool, initially written by Mr. Kasprzak for use at UB. We have decided to incorporate it as an integral part of the present volume. Appendix A is dedicated to problems which are solved by the use of these vehicle dynamics tools, and readers are encouraged to explore and experiment with these programs to get a better feel for material presented in RCVD. Although the programs are simplistic in relation to more elaborate vehicle dynamics models, they are very useful in giving rapid, first-order solutions to many practical problems.

We were fortunate in having the manuscript of this book proofed by an outstanding vehicle dynamicist—Donald Van Dis, Chrysler Proving Grounds (retired). Proofing and copy editing was also done by SAE, however, the authors are responsible for the accuracy of the material in the book and on the CD. As with our previous books, we welcome comments and corrections. Please email to the contact address at www.millikenresearch.com or mail to the SAE Publications Group. As errors are found, an errata page will be posted on our website.

The following paragraphs are taken directly from the "General Introduction" to the original *RCVD Workbook*:

The reaction to *Race Car Vehicle Dynamics* has been very gratifying. Readers include professional race engineers and vehicle dynamicists, amateur racers and enthusiasts. As we said in the Preface, our original intent was:

> ...to make available to the racing community, and race engineers in particular, an understandable summary of vehicle dynamics technology as it has developed over the last 60 years. This technology provides a consistent way of looking at the vehicle part of handling problems. We have tried to follow a path between a "theoretical" textbook on vehicle dynamics (which could miss the unique needs of racers) and producing a "popular" book on handling (skipping over the engineering details). While much of the material is mathematical, the practical application of the theory is becoming available in computer programs.

In keeping with the original objective of the book, we used algebra and simple mechanics wherever possible—avoiding higher mathematics.

Since publication we have been pleasantly surprised that the book is being used as a formal text in the university environment. There are two major interests: as an introduction to the general area of vehicle dynamics and as a resource to those students engaged in competition, notably the various SAE-sponsored series. These two uses can apply at both the undergraduate and graduate level where there is no need to restrict a course to simple mathematics.

In the academic environment, problems have traditionally been included to challenge the student. This supplement includes problems, experiments and additional material and references for those who wish to become more proficient in vehicle dynamics analysis. In general, the problems are geared to a course specifically for third- or fourth-level undergraduate or graduate-level students. A course in vehicle dynamics will be inherently multi-disciplinary, as opposed to a typical engineering course that goes in depth into a particular engineering topic. The justification for organizing a course around the broad range of topics covered by vehicle dynamics is the general interest that exists in automobiles and race cars.

For those who wish to get the most out of such a course or have in mind becoming a professional vehicle dynamicist, there are a number of prerequisite studies. In *Race Car Vehicle Dynamics*, we had originally intended to include summaries of a number of related disciplines such as engineering mechanics, dynamics, control theory and vibration. Instead, we finally recognized that these general subjects are much better treated by existing established texts and we have provided references to them in this supplement. We believe these will also prove useful to those in the race car business who do not have these prerequisites from a college background but are interested in self-study and continuing education.

In the real world of racing there is no shortage of problems. We encourage those actively campaigning or building a race car to use these problems as a guide, perhaps substituting numbers for their own car, or creating their own problems from reality.

Doug Milliken
Edward Kasprzak
Daniel Metz
Bill Milliken

April 2003

CHAPTER 1

The Problem Imposed by Racing

Solutions to this chapter's problems begin on page 2.

1.1 Problems

1. Assume a car with a 1g lateral acceleration capability is cornering on a skidpad with constant radius R. Plot vehicle speed versus R at this acceleration for radii between 50 ft. and 1000 ft.

2. A certain (fictitious) race car has a circular "g-g" diagram with a radius of 1g. Predict this car's lap time and average speed on a 1.0 mile oval consisting of two 1/4 mile straights connecting two 1/4 mile constant radius turns. For simplicity, assume the transition from straightaway to corner and vice versa happens instantaneously.

3. The previous two questions assumed a circular "g-g" diagram boundary. Identify at least two ways in which this simplification varies from real vehicles.

4. Derive the lateral acceleration formula $A_Y = V^2/R$ from RCVD[1] Figure 1.3, p.6.

[1]RCVD is an abbreviation for *Race Car Vehicle Dynamics*. It is used throughout this book.

5. a. In your own words, formulate one or more problems in motor racing and/or vehicle dynamics that are of interest to you.

   b. As you complete your course of study, write an answer for each of the problems you generated in Part a.

1.2 Simple Measurements and Experiments

ATTENTION: All experiments should be conducted at legal speeds and under safe conditions.

1. Equip your vehicle with a g·Analyst, onboard data acquisition system or other accelerometer. Drive on a constant radius curve (such as a coned-circle on a parking lot, skidpad or expressway cloverleaf) at a constant speed. Use the speed and lateral acceleration to determine the radius. Repeat at different speeds on the same curve. Do you get the same answer? Why or why not?

1.3 Problem Answers

1. Rearranging $A_Y = V^2/R$ gives $V = \sqrt{A_Y R}$ with A_Y in ft./sec.^{2}, R in feet and V in ft./sec. In more common units:

$$V = \frac{22}{15}\sqrt{32.2 A_Y R}$$

   with A_Y in g-units, R in feet and V in mph. Figure 1.1 plots the result of this equation at $A_Y = 1\text{g}$.

2. First, determine the radius of the corners.

$$C = \pi d = \pi(2R)$$

$$2\left(\frac{1}{4}\text{mile}\right) = 2\pi R$$

$$R = 0.0795 \text{ mile} = 420.2 \text{ ft.}$$

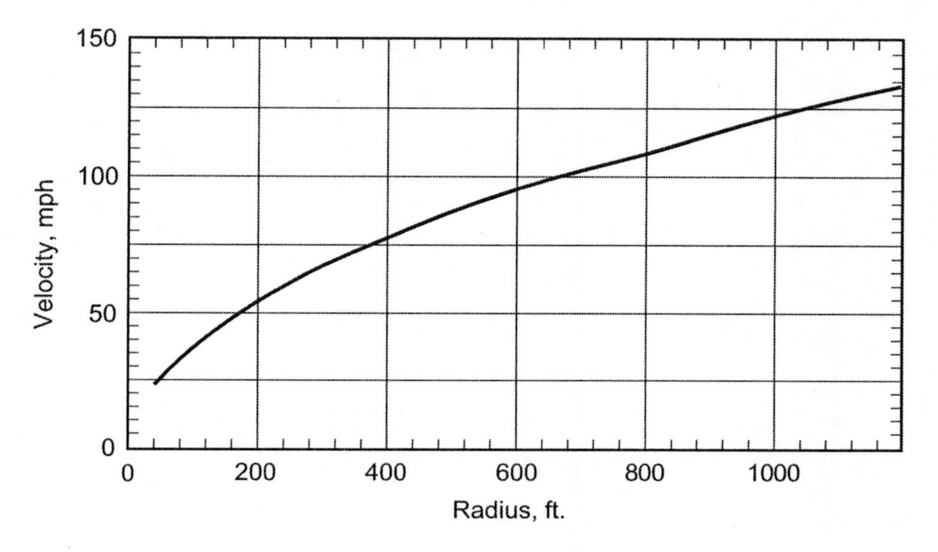

Figure 1.1 *Cornering speeds at 1g lateral acceleration*

We know the vehicle has $A_Y = 1g$, so we can calculate the speed in the corners:

$$V = \sqrt{32.2 \times 420.2} = 116.3 \text{ ft./sec.} = 79.3 \text{ mph}$$

This determines the speed at the start and end of the straightaways. On the straights, the vehicle accelerates at 1g and decelerates at 1g. Since the acceleration equals the deceleration and the initial speed equals the final speed, this vehicle transitions from accelerating to braking at the midpoint of the straight. Acceleration and braking each take place over 1/8 mile, or 660 ft. Using a formula from high-school physics:

$$V_f^2 = V_i^2 + 2a\,(\Delta\text{distance})$$

$$V_f = \sqrt{116.3^2 + 2\,(32.2)\,(660)} = 236.7 \text{ ft./sec.} = 161.4 \text{ mph}$$

This is the maximum speed on the straight. Since speed varies linearly with constant acceleration, the average speed on the straights can be calculated from:

$$V_{avg} = \frac{116.3 + 236.7}{2} = 176.5 \text{ ft./sec.} = 120.3 \text{ mph}$$

The time to complete one lap can now be calculated:

$$t = \left.\frac{\text{distance}}{\text{speed}}\right|_{\text{straights}} + \left.\frac{\text{distance}}{\text{speed}}\right|_{\text{corners}}$$

$$t = \frac{2\,(1320\text{ ft.})}{176.5\text{ ft./sec.}} + \frac{2\,(1320\text{ ft.})}{116.3\text{ ft./sec.}} = 37.65\text{ sec.}$$

This results in an average speed of:

$$V_{avg} = \frac{5280\text{ ft.}}{28.39\text{ sec.}} = 140.2\text{ ft./sec.} = 95.6\text{ mph}$$

Note: This calculation is in the same spirit as the one performed at CAL in 1971 for the enlargement of the Watkins Glen circuit, mentioned on RCVD p.341, which used a point mass vehicle model with a "g-g" acceleration boundary. The CAL Watkins Glen calculations, however, included smooth transitions into and out of the corners.

3. The "g-g" diagram is rarely circular. Available power routinely truncates the diagram in longitudinal acceleration as seen in RCVD Figure 1.5, p.10. Peak lateral acceleration, longitudinal acceleration and longitudinal deceleration typically have different values due to tire properties, weight distribution, weight transfer, etc. The "g-g" diagram boundaries may also be speed dependent due primarily to aerodynamics.

4. Start by redrawing RCVD Figure 1.3 in terms of vectors occurring at two different instants (denoted with subscripts 1 and 2) as shown in Figure 1.2. At each instant the radius and velocity have the same magnitudes, denoted R and V, respectively. Note, however, that they have different directions. This gives rise to the centripetal acceleration.

   $\overrightarrow{R_1}$ is perpendicular to $\overrightarrow{V_1}$ and $\overrightarrow{R_2}$ is perpendicular to $\overrightarrow{V_2}$. $\overrightarrow{R_2}$ minus $\overrightarrow{R_1}$ gives $\overrightarrow{\Delta R}$, the change in radius vector between these two instants. Similarly, $\overrightarrow{V_2}$ minus $\overrightarrow{V_1}$ gives the change in velocity vector $\overrightarrow{\Delta V}$. The following can be stated:

$$\frac{\left|\overrightarrow{\Delta V}\right|}{V} = \frac{\left|\overrightarrow{\Delta R}\right|}{R} \Rightarrow \left|\overrightarrow{\Delta V}\right| = \frac{V}{R}\left|\overrightarrow{\Delta R}\right|$$

   Instantaneous acceleration is defined as the change in velocity with respect to time

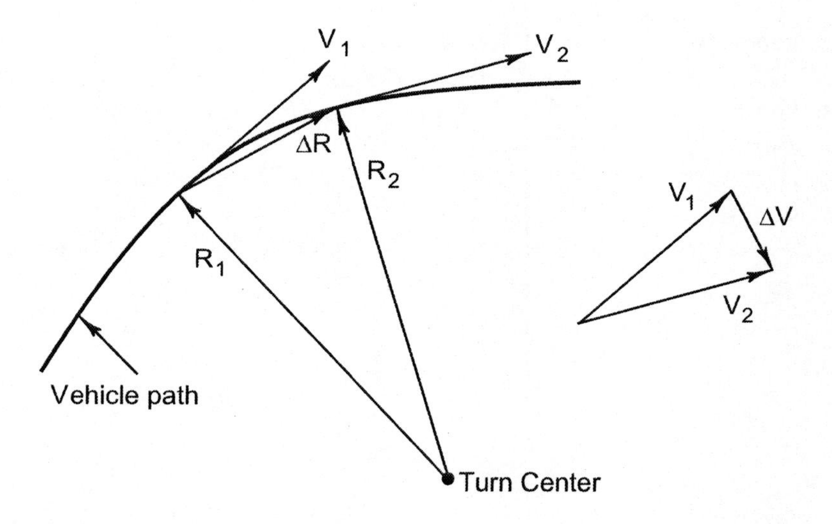

Figure 1.2 *Vector diagram for the calculation*

as the time increment approaches zero.

$$a_y \equiv \lim_{\Delta t \to 0} \frac{\left|\overrightarrow{\Delta V}\right|}{\Delta t}$$

Combining the above two expressions:

$$a_y \doteq \lim_{\Delta t \to 0} \frac{(V/R)\,|\Delta R|}{\Delta t} = \frac{V}{R} \lim_{\Delta t \to 0} \frac{|\Delta R|}{\Delta t} = \frac{V}{R} V = \frac{V^2}{R}$$

5. Some problems in vehicle dynamics and motor racing have been solved already, while others have not. Problems and solutions vary widely in complexity, and there's often more than one solution to a given problem. It's part of the allure of the discipline. Through a series of problems and their solutions this book aims to help deepen the reader's understanding of vehicle dynamics.

1.4 Comments on Simple Measurements and Experiments

1. Ideally you *should* calculate the same radius at any speed from vehicle speed and lateral acceleration, but several factors can prevent this from happening. First, it is very difficult to drive at a constant speed and constant steer angle on a

given radius—your data will invariably have some amount of driver-induced noise. Even if driven perfectly the sensors will still be prone to noise and calibration errors. How do you know the curve you're driving on is a constant radius to begin with? Is the road banked? These last two can be eliminated by driving a measured circle on a flat surface like a parking lot. There are other instrumentation issues as well. Body roll affects the accelerometer, and a speedometer calibrated on a straight, measured road will need to be recalibrated for a curved path.

CHAPTER **2**

Tire Behavior

Solutions to this chapter's problems begin on page 13.

2.1 Problems

1. Calculate stopping distance from 62 mph for a vehicle with a tire/road friction coefficient equal to 0.6, 0.8 and 1.0, successively. Assume the vehicle can be modeled using particle dynamics. Also, calculate the car's maximum forward velocity for each coefficient when driven continuously around a 100 ft. radius skidpad.

2. The Eagle ZR tire shown in RCVD Figure 2.44, p.78, is used on a Corvette with a test weight of 3500 lb. and having a 52/48 weight distribution. Use the cornering stiffnesses (i.e., a linearized tire) and the actual curves themselves (a nonlinear tire) to calculate the cornering force at a slip angle of 1.5 deg. on the front and rear tires. What is the percent error between linearized and nonlinear tire models? Repeat the calculation at 4.0 deg. Interpolation between load curves will be necessary. For simplicity, ignore weight transfer.

3. Consider the Street Corvette P275/40 ZR17 tires of RCVD Figure 2.44, p.78. Calculate the total drag of a four-wheel, 3000 lb. vehicle if all tires are operating at a slip angle of 3 deg. and have a rolling resistance coefficient of 0.022. Calculate the

power required to drive this vehicle at 30 mph, ignoring aerodynamic forces. For simplicity, assume all tires have equal normal loads.

4. A car with P215/60-R15 Goodyear Eagle GT-S tires (RCVD Figure 2.42, p.76) is cornering. Each front wheel is operating at 3 deg. of slip angle. The inside front wheel has a normal load of −900 lb. and the outside front wheel has a normal load of −1350 lb. Furthermore, the kingpin has no offset and the caster angle results in 1.125 in. of mechanical trail. Calculate the torque about the kingpin for each front wheel.

5. Qualitatively characterize the differences between the Goodyear P275/40-ZR17 Eagle street tire shown in RCVD Figure 2.44, p.78, and the Goodyear IndyCar road course 27.0x14.5-15 rear tire shown in RCVD Figure 2.45, p.79. Consider the following characteristics:

   a. Which tire has the highest tire/road coefficient of friction?

   b. Which tire has the highest cornering stiffness?

   c. Which tire has the most aligning torque stiffness (normalized per unit normal load)?

   d. Which tire would be most "forgiving" (feel free to experiment with your own definition of "forgiving" here)?

6. Make a photocopy of the tire data presented in RCVD Figure 2.8, p.26 and assume that the tire behaves according to classic friction *circle* principles (not a friction ellipse). On your photocopy of RCVD Figure 2.8, use friction circle calculations to superimpose a family of curves showing available longitudinal forces at various slip angles and loads. For this tire, what is the maximum amount of longitudinal force available when the tire is being cornered with a slip angle of 4 deg. at 1350 lb. normal load?

7. Using the tire data shown in RCVD Figure 2.46, p.80, for a Goodyear Formula One front tire, normalize both slip angle versus lateral force and slip angle versus aligning torque per unit normal load F_Z. Make a table of the following for each normal load. Are the results as you would expect? Why (not)?

   a. Peak Friction Coefficient

   b. Cornering Coefficient (Cornering Stiffness / Normal Load)

   c. Aligning Torque Coefficient (Aligning Torque Stiffness / Normal Load)

8. Assume you are the manager of the TIRF facility shown in RCVD Figure 2.6, p.24. Briefly describe a test program which would completely characterize a given tire.

9. Suppose a vehicle had been originally equipped with the Eagle GT-S tires shown in RCVD Figure 2.42, p.76. Qualitatively characterize the expected changes in vehicle response to a small step steering input (within the linear range of tire performance) if:

   a. The tire pressure is raised by 25%.

   b. The tire pressure is lowered by 25%.

   c. The tires are changed to those shown in RCVD Figure 2.43, p.77.

   In answering parts a. and b. comment on speed of response and maximum cornering capability. Additionally for part c., comment on the maximum lateral acceleration performance with the two sets of tires.

10. [1]Many vehicle dynamics problems require knowledge of the tire loaded radius and tire spring rate. These vary with operating conditions such as load, speed and lateral force. Figures 2.1 and 2.2 provide some data on these parameters.

    To get a feel for these tire effects, produce analytic models (curve fits or regressions) for the following cases:

    a. Loaded radius vs. lateral force

    b. Loaded radius vs. load and speed

    c. Tire spring rate vs. load

11. [2]Tires do not generate lateral force instantaneously in the presence of a slip angle. Instead, the tire needs to roll through a certain distance before full, steady-state lateral force is achieved. One way to model this is through the use of tire "relaxation length". Relaxation length is typically 0.5 to 1.0 tire circumference, independent of speed. The build-up of lateral force can be modeled as a first-order system where the time constant is given by the time required for the tire to roll through a distance of one relaxation length.

---

[1]This problem suggested by Dr. Alex Moulton.
[2]This problem suggested by Bangalore Suresh.

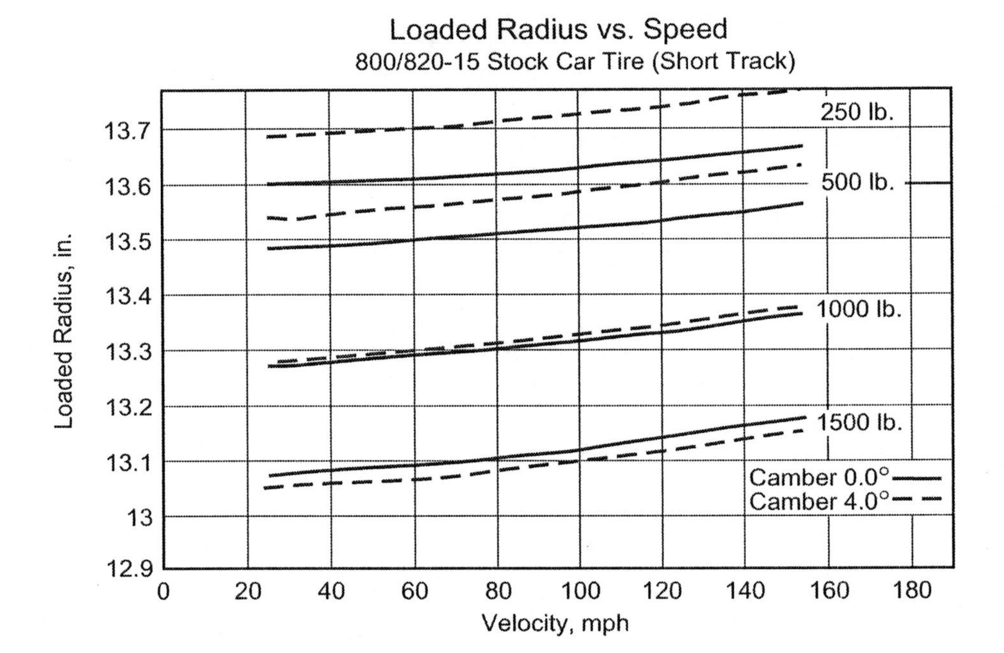

Figure 2.1 *Goodyear data—Short Track Stock*

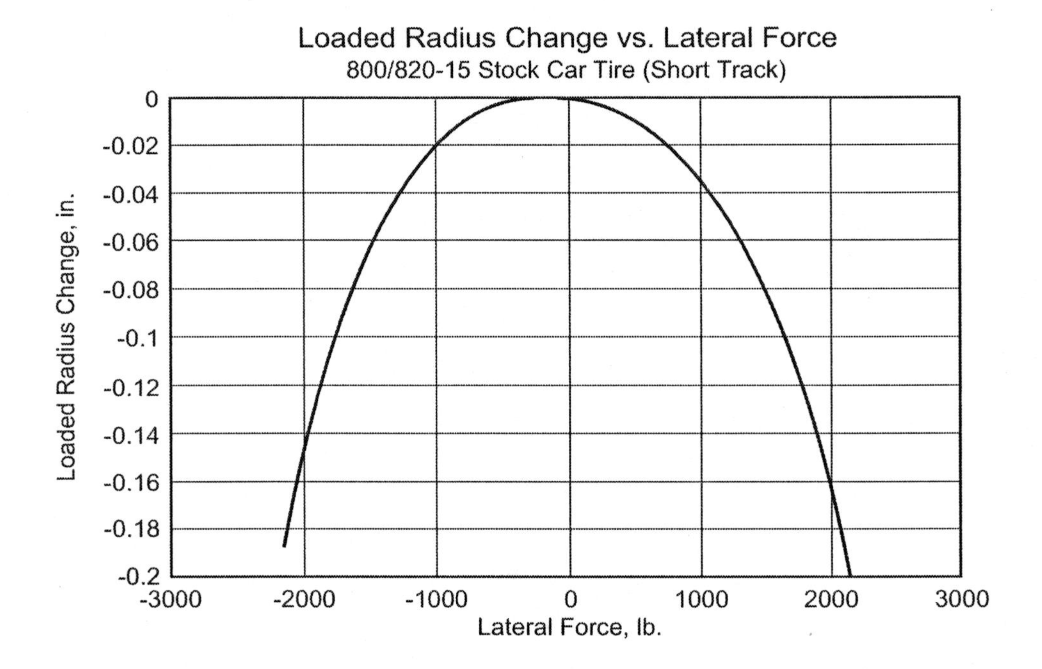

Figure 2.2 *Goodyear data, valid for all loads*

For a tire with a 6 ft. circumference and a relaxation length equal to one circumference, perform the following tasks:

a. Calculate the time constant as a function of speed.

b. Determine the lateral force response of the front tires to a step change in slip angle. Plot the response for a tire with cornering stiffness $C_\alpha = -335$ lb./deg. and a step change to a slip angle of -2 deg.

c. Determine the lateral force response to a ramp-steer input, an input typical of ordinary driving. Assume a ramp steer at the roadwheel of 5 deg./sec. (or 70 deg./sec. at the steering wheel with a 14:1 steering ratio), the tire properties from Question 11b and linear range behavior. Plot the response with and without tire lag at 30 mph for 2 seconds of ramp input.

d. Comment on the significance of tire lag to overall vehicle response at low speeds and also at racing speeds.

12. The relaxation length concept used in the previous problem is one way to model transient tire forces. An alternative technique makes use of tire lateral stiffness, k_L. This is used in Hooke's Law to give the lateral force:

$$F_y = k_L \times y$$

where y is the lateral displacement of the contact patch relative to the wheel center.

a. Show that the above equation leads to the following differential equation where u is the longitudinal velocity, C is the cornering stiffness and $\delta(t)$ is the wheel steer angle as a function of time:

$$\dot{y} - \frac{k_L u}{C} y = -u \times \delta(t)$$

Hint: Assume linear range tire behavior.

b. Solve the differential equation in Question 12a for a 1 deg. step input.

c. Solve the differential equation in Question 12a for a sinusoidal input of the form $A \sin(\omega t)$.

d. Create a magnitude Bode plot of the response to a sinusoidal input. Compare your solution with the results presented by Weber and Persch[3].

---

[3]SAE Paper No. 760030 *Frequency Response of Tires: Slip Angle and Lateral Force*, abstract on CD.

2.2 Simple Measurements and Experiments

> NOTE: All experiments should be carried out at low speeds (say, below 30 mph) under safe conditions. A shopping market or campus parking lot on early Sunday morning often makes a good skidpad.
>
> WARNING: The following experiments can produce significant amounts of tire wear in the shoulder area of the tires. Do not continue the experiments for many laps unless you are unconcerned about tire wear. Tire wear can be reduced by performing the experiments on wet or snow covered surfaces.

1. Use a step steering input to enter and drive around a constant radius skidpad (suggested radius approximately 50 ft.). Examine and qualitatively comment on the changes in car behavior which can be observed when:

   a. The front tire pressures are increased by 10 psi over stock.

   b. The rear tire pressures are increased by 10 psi over stock.

   c. Both front and rear tire pressures are increased by 10 psi over stock.

   d. Both front and rear tire pressures are decreased by 10 psi under stock.

   Characteristics to comment on include turn in, rapidity of response, ability to hold the constant radius turn at various forward velocities and car "stability" (the question of "stability" is addressed in Chapter 5).

2. Using an (approximately) sinusoidal steering input and two speeds (walking speed and perhaps 30 mph), provide steering wheel inputs of about ±45 deg. at various frequencies and qualitatively comment on the car's behavior.

3. For the tires on your own vehicle, measure free radius, loaded radius and effective radius (effective radius, R_e = distance traveled in one revolution divided by 2π) at approximately zero speed. Compare these radii and order them in order of increasing magnitude. Would you expect centrifugal effects to alter these three radii? If not, why not? If so, would you expect them to increase or decrease?

   Also, present an experiment that could be used to measure the tire contact patch print shape and area.

4. Conduct a coastdown test[4] to estimate the total rolling resistance of your vehicle in terms of an overall rolling resistance coefficient (normalized by car weight, of course). Comment on the effects of aerodynamic drag, headwind/tailwind and/or the presence of a grade or roadway crown on such a test. Consider towing the car with a fish scale (spring scale). What problems would you expect if you tried to measure rolling resistance with a fish scale? How about with a strain-gauged, rigid tow bar?

2.3 Problem Answers

1. Assume the vehicle is a point mass and start with two formulas for a particle experiencing constant acceleration.

$$s_f = s_0 + v_0 t + \frac{1}{2} a t^2 \quad \text{and} \quad v_f = v_0 + at$$

In this case we have:

$$s_f = 0 + \frac{22 \text{ ft./sec.}}{15 \text{ mph}} (62 \text{ mph})\, t + \frac{1}{2}\left(32.2 \text{ ft./sec.}^2\right) \mu t^2$$

$$v_f = 0 = \frac{22 \text{ ft./sec.}}{15 \text{ mph}} (62 \text{ mph}) + \left(32.2 \text{ ft./sec.}^2\right) \mu t$$

Combining and solving for stopping distance s_f:

$$s_f = \frac{V_0^2}{30\mu}$$

Substituting friction coefficient values yields the following stopping distances:

Friction Coeff., μ –	Stopping Dist., s_f ft.
0.6	213.6
0.8	160.2
1.0	128.1

[4]You may want to review SAE Recommended Practice J1263, which describes coastdown test procedures, before proceeding with this experiment.

On the skidpad, with radius R in ft. and speed V in ft./sec.:

$$a = \frac{V^2}{R} \Rightarrow V = \sqrt{32.2\mu R}$$

The following speeds result on a 100 ft. skidpad:

μ	V	
–	ft./sec.	mph
0.6	44.0	30.0
0.8	50.8	34.6
1.0	56.7	38.7

2. The front of the vehicle weighs $0.52 \times 3500 = 1820$ lb. and (ignoring weight transfer) each front wheel carries half this load, or 910 lb. The rear of the vehicle weighs $0.48 \times 3500 = 1680$ lb. and, again ignoring weight transfer, each rear wheel carries 840 lb.

The following lateral force values can be read off RCVD Figure 2.44:

Lateral Force (lb.)

F_Z	α (deg.)		
lb.	0.5	1.5	4.0
405	−75	−205	−380
920	−155	−430	−855

Since the front wheels each have a normal load of 910 lb. and the figure gives a curve at 920 lb., we will avoid interpolation by using the 920 lb. values. The value at 0.5 deg. is used to determine the cornering stiffness. At 920 lb. the cornering stiffness is −310 lb./deg. This value multiplied by the slip angle generates the linear value for lateral force.

Results for the front tires:

α	Lateral Force		Error
	linear	nonlinear	
deg.	lb.	lb.	%
1.5	−465	−430	8.1
4.0	−1240	−855	45.0

Interpolation is required for the rear tires. The lateral force at 840 lb. and 0.5 deg. slip angle can be calculated from:

$$-155 - \left(\frac{920-840}{920-405}\right)(-75) = -143.3$$

A similar calculation for the other two slip angles results in the following table:

Lateral Force (lb.)

F_Z	α (deg.)		
lb.	0.5	1.5	4.0
840	−143.3	−398.1	−796.0

At 840 lb. the cornering stiffness is −186.6 lb./deg. We are now ready to make the comparison for the rear tires:

α	Lateral Force		Error
	linear	nonlinear	
deg.	lb.	lb.	%
1.5	-429.9	-398.1	8.0
4.0	-1146.4	-796.0	44.0

These calculations show that approximating lateral force using cornering stiffness is valid for small slip angles (the linear range) but the percent error increases significantly once the tire is operating in the transitional range. Repeating this comparison at, say, 6 deg. would result in over 100% error.

3. Each wheel has a normal load of $3000/4 = 750$ lb. and is operating at 3 deg. of slip angle. Interpolation from RCVD Figure 2.44 gives a lateral force of −580 lb. at this operating condition for each tire, a total of −2320 lb. This causes an induced drag of $-2320 \times \sin(3\text{ deg.}) = -121.4$ lb. opposing vehicle motion.

   Rolling resistance is calculated as $3000 \times 0.022 = 66$ lb. opposing vehicle motion. Adding this to the induced drag gives 187.4 lb. of total drag force.

   Since Power = Force × Velocity we can calculate the horsepower required to overcome the tire drag:

$$\text{Power} = 187.4\text{ lb.} \times 44\text{ ft./sec.} = 8245.6\,\frac{\text{lb.-ft.}}{\text{sec.}} = 15.0\text{ hp}$$

   Thus, it takes 15 horsepower to overcome the rolling resistance and induced drag with all tires operating at 3 deg. of slip angle.

4. First the inside tire. At −900 lb. normal load and 3 deg. of slip angle this tire produces −800 lb. of lateral force and 40 lb.-ft. of aligning torque. The lateral force acts on a moment arm equal to the mechanical trail which, when combined with the aligning torque, gives a total kingpin moment of:

$$-800 \text{ lb.} \times \left(-1.125/12\right) \text{ ft.} + 40 \text{ lb.-ft.} = 115 \text{ lb.-ft.}$$

The outside wheel produces a lateral force of −1050 lb. and an aligning torque of 105 lb.-ft. at 3 deg. of slip angle and −1350 lb. of normal load. A similar calculation gives the kingpin torque:

$$-1050 \text{ lb.} \times \left(-1.125/12\right) \text{ ft.} + 105 \text{ lb.-ft.} = 203 \text{ lb.-ft.}$$

The sum of the inside and outside kingpin torques are passed through the linkages, rack and steering column to the steering wheel. Note that, since the kingpin is inclined, the above answers are only approximately correct. To be complete the above answers would need to be multiplied by the cosine of the caster angle, although the difference in this case is very small.

5. a. Friction coefficient is measured at the peak of the lateral force versus slip angle curve according to the following formula:

$$\mu = F_{y_{peak}} / F_z$$

For the Eagle street tire:

F_z lb.	$F_{y_{peak}}$ lb.	μ –
405	450	1.11
920	950	1.03
1450	1400	0.97
1956	1850	0.95

For the IndyCar tire:

F_z lb.	$F_{y_{peak}}$ lb.	μ –
1000	1800	1.80
1400	2300	1.64
1800	2550	1.42

The IndyCar tire has a higher friction coefficient. Note that for both tires, friction coefficient decreases with increasing tire load. This is termed "tire load sensitivity".

b. Answers are based on the highest tested load for each tire. The street Eagle, at a load of 1956 lb., has a cornering stiffness of:

$$C = \left.\frac{F_y}{\alpha}\right|_{\alpha=0} \approx \frac{-2000}{3.6} = -556 \text{ lb./deg.}$$

The IndyCar tire, at a load of 1800 lb., has a cornering stiffness of:

$$C = \left.\frac{F_y}{\alpha}\right|_{\alpha=0} \approx \frac{-3000}{3.4} = -882 \text{ lb./deg.}$$

The IndyCar tire has a higher cornering stiffness.

c. Use the following formula to calculate the aligning torque stiffness normalized by normal load (which is termed "aligning torque coefficient"):

$$M_Z' = \left(\left.\frac{\Delta M_Z}{\Delta\alpha}\right|_{\alpha=0}\right) \Big/ F_Z$$

For the Eagle street tire:

F_Z lb.	M_Z' –
405	0.020
920	0.025
1450	0.033
1956	0.035

For the IndyCar tire:

F_Z lb.	M_Z' –
1000	0.145
1400	0.161
1800	0.153

The IndyCar tire has more aligning torque and aligning torque stiffness per unit normal load than the street Eagle.

d. In vehicle dynamics terms, a "forgiving" tire is usually thought of as one which has a long and smooth transitional region. Sketches of a forgiving and unforgiving tire are shown in Figure 2.3.

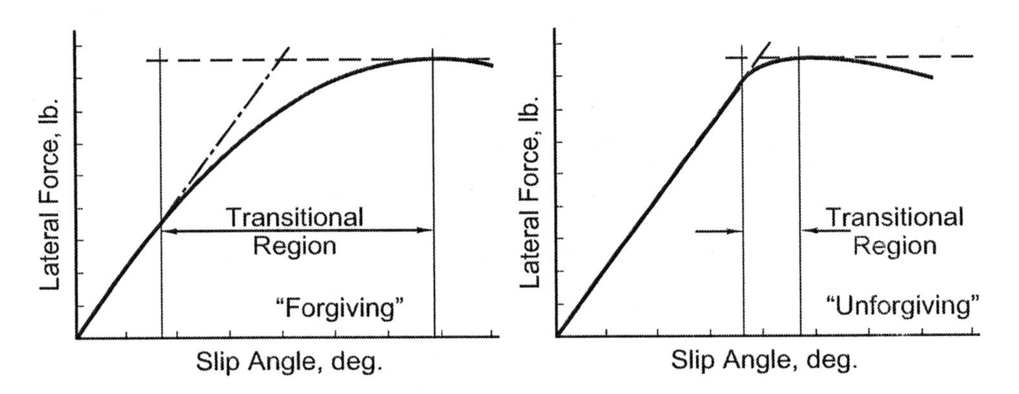

Figure 2.3 *A "forgiving" and an "unforgiving" tire*

The Eagle and IndyCar tires have similar transitional ranges. We would expect them to be more-or-less equally forgiving.

In conclusion, the two tires are surprisingly similar. The tires exhibit the same characteristics, although the IndyCar tire does so at a higher performance level.

6. The friction *circle* model assumes that peak lateral friction equals peak longitudinal friction at each load. Thus, at the peak of the lateral force versus slip angle curve the available longitudinal force must be zero. Similarly, at zero lateral force the longitudinal force is equal to the load multiplied by the friction coefficient.

   Between these two points the available lateral and longitudinal forces sum vectorally to the load times the friction coefficient. That is,

$$F_X^2 + F_Y^2 = (\mu F_Z)^2$$

   This can be solved for F_X to determine the available longitudinal force:

$$F_X = \sqrt{(\mu F_Z)^2 - F_Y^2}$$

   Applying this to the lateral force data in the table results in the curves shown in Figure 2.4.

7. a. To calculate peak friction coefficient, find the maximum lateral force on a

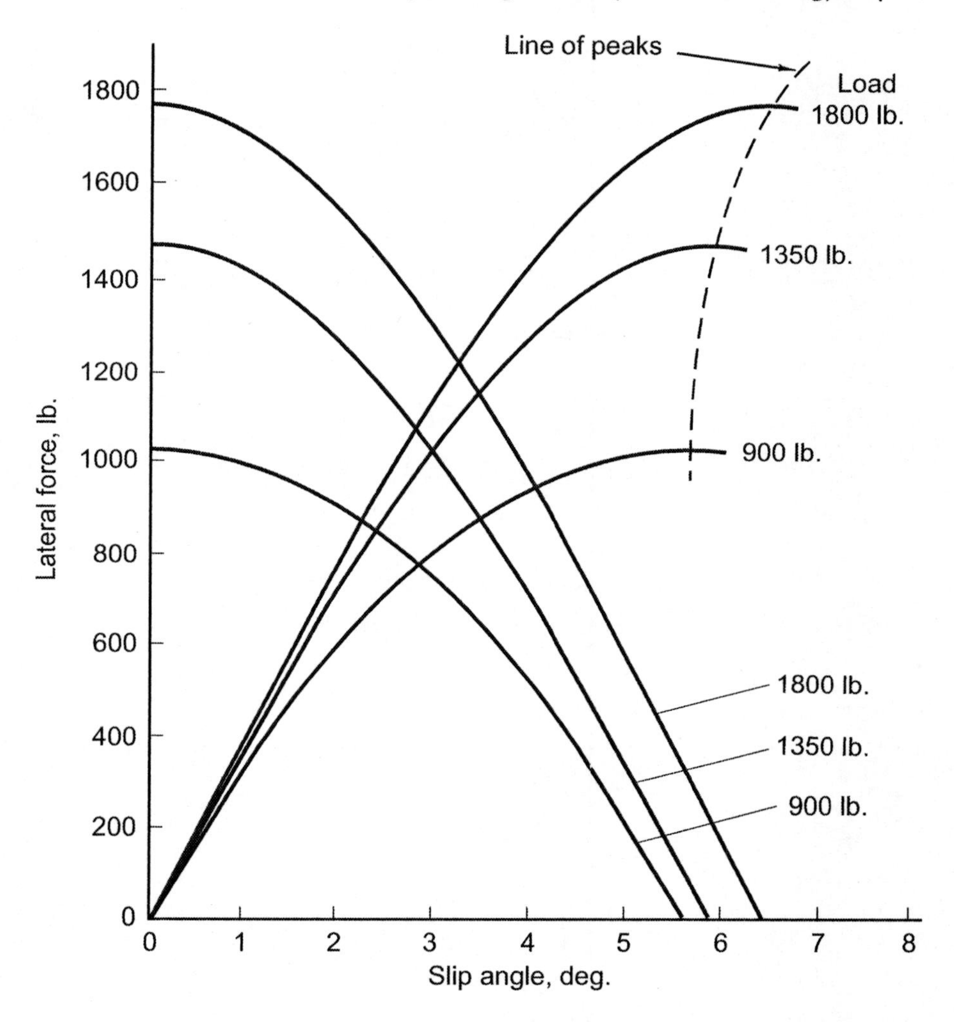

***Figure 2.4** RCVD Figure 2.8 with lines of available longitudinal force*

given load curve and divide by the normal load, $\mu = F_{y_{peak}} / F_Z$. See the table following Answer 7c.

b. To calculate cornering coefficient, determine the cornering stiffness for each load by measuring the slope of the lateral force curve at zero slip angle. Then divide by the normal load. See table following Answer 7c.

c. To calculate aligning torque coefficient, measure the initial slope of the aligning torque versus slip angle curve for a given load and then divide by the load.

Results are given in the following table:

Load lb.	Friction Coeff. –	Cornering Coeff. –	Aligning Coeff. –
400	1.75	1.17	0.31
600	1.70	0.96	0.13
800	1.50	0.89	0.08
1000	1.40	0.71	0.02

The results are as expected. Friction coefficient decreases with increasing normal load, reflecting the tire load sensitivity inherent to all tires. Similarly, cornering coefficient and aligning torque coefficient decrease with increasing normal load.

8. There is no unique answer to this problem. Variables of interest in a tire test program would be:

   - Range of slip angles sufficient to find the peak at each normal load
   - Range of normal loads from near-zero to above the rated tire load
   - Range of camber angles sufficient to cover the expected operating range
   - Range of forward velocities from near-zero to the maximum tire rated speed
   - Range of slip ratios sufficient to determine the peak longitudinal force in driving and braking
   - Range of test belt surfaces, if more than one is available
   - Range of tire pressures

   Numerical values for the above ranges depend on the tire and its application. Street tires, race tires and truck tires would have very different bounds. Additionally, the design of a good tire test involves more than just determining the right ranges. The test sequence is very important and must consider tire temperatures and tire wear.

   Since full characterization of a tire in this manner is such a large task it is rarely attempted. Instead, a suitable subset of the full test is conducted, focusing on areas of interest. There are, however, ways to characterize a tire without such a comprehensive test—see Nondimensional Tire Theory in Chapter 14.

9. a. As indicated on RCVD p.54, increasing tire pressure increases tire stiffness. Therefore, raising tire pressures by 25% would produce a more rapid turn-in and higher steady-state cornering acceleration per unit of steering wheel input. (This is sometimes called control gain or steering sensitivity. See RCVD Chapter 5 or SAE J670e, Definition 9.2.7.)

b. Reducing tire pressures reduces tire sensitivity and control gain.

c. If the tires are changed to those shown in RCVD Figure 2.43, p.77, control gain would be increased because these tires are inherently much stiffer than those in RCVD Figure 2.42 (to see this, estimate the cornering stiffness for each tire).

The tires of RCVD Figure 2.43 have considerably more aligning torque (all other things being equal) than those of RCVD Figure 2.42. To estimate maximum lateral cornering capability, we need to estimate the tire/road friction coefficient. At their lightest loads, respectively, the tires in RCVD Figures 2.42 and 2.43 have friction coefficients of 1.17 and 1.50. The tire of RCVD Figure 2.43 gives a higher cornering capability.

10. a. This loaded radius versus lateral force data can be fit quite well with an offset parabola:

$$\Delta R_\ell = \left(3.5 \times 10^{-8}\right)\left(F_y + 150\right)^2$$

Of course, other fits could be used and give acceptable results. If you assume the tire should be symmetrical, the left and right side data could be averaged before applying a regression.

b. Figure 2.1 provides the data needed to mathematically model loaded radius versus load and speed. Consider the 0 deg. camber condition. The following linear fits apply for each load where V is the speed in mph:

Normal Load lb.	Loaded Radius in.
250	$13.58 + 0.00049V$
500	$13.46 + 0.00061V$
1000	$13.25 + 0.00078V$
1500	$13.05 + 0.00083V$

We can take this one step further by introducing the normal load into the equation. This equation now applies for all data at 0 deg. camber:

$$R_\ell = 13.65 - 0.0004F_Z + \left(2.6 \times 10^{-4} + 8.4 \times 10^{-7}F_Z - 3.2 \times 10^{-10}F_Z^2\right)V$$

Coefficients for the above expression were calculated using MATLAB. The process could be repeated at 4 deg. camber.

c. Tire spring rate is the slope of the normal load versus deflection curve. Plotting data at 60 and 120 mph produces Figure 2.5.

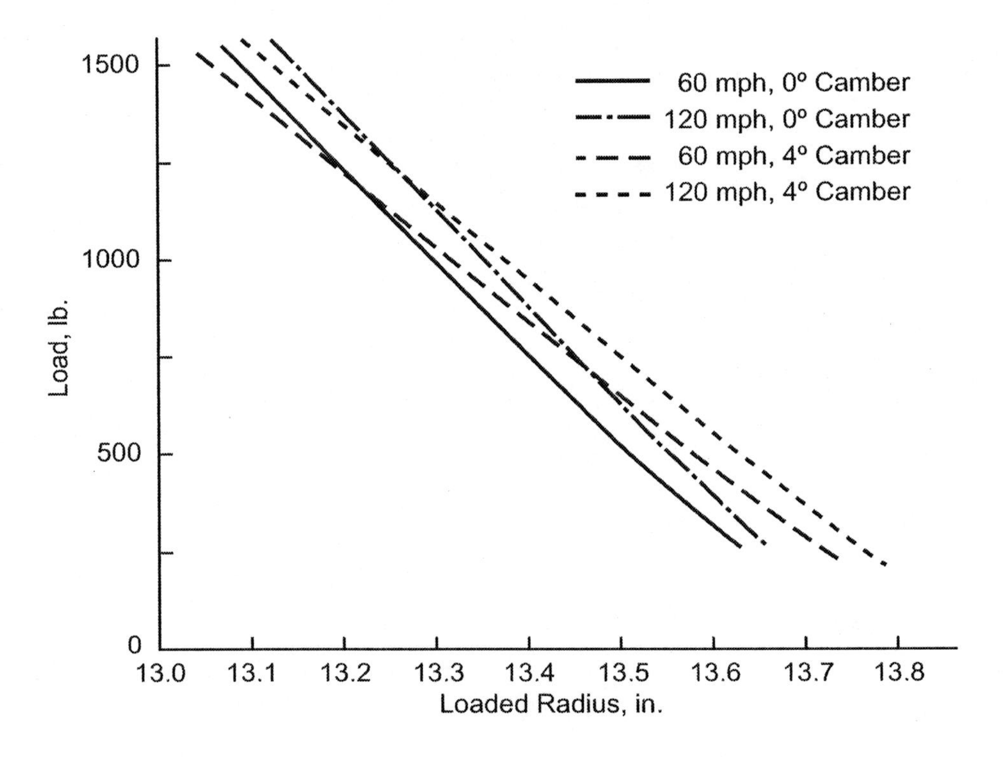

Figure 2.5 *Graph to determine spring rate*

From which we can extract these spring rates:

Speed mph	Camber deg.	Spring Rate lb./in.
60	0	127.6
120	0	145.2
60	4	108.9
120	4	109.1

For this tire, spring rate is load-independent, decreases with increasing camber and increases with increasing speed. By extracting additional spring rates at other speeds, enough information could be collected to fit a polynomial or a response surface to the data.

11. a. [5]With the relaxation length equal to the circumference, the time constant is calculated by:

$$\tau = \frac{\text{Relaxation Length}}{\text{Speed}} = \frac{C}{V}$$

For a tire with a circumference (and relaxation length) of 6 ft. the time constant varies with speed according to Figure 2.6. Also shown are curves for when the relaxation length is a fraction of the circumference.

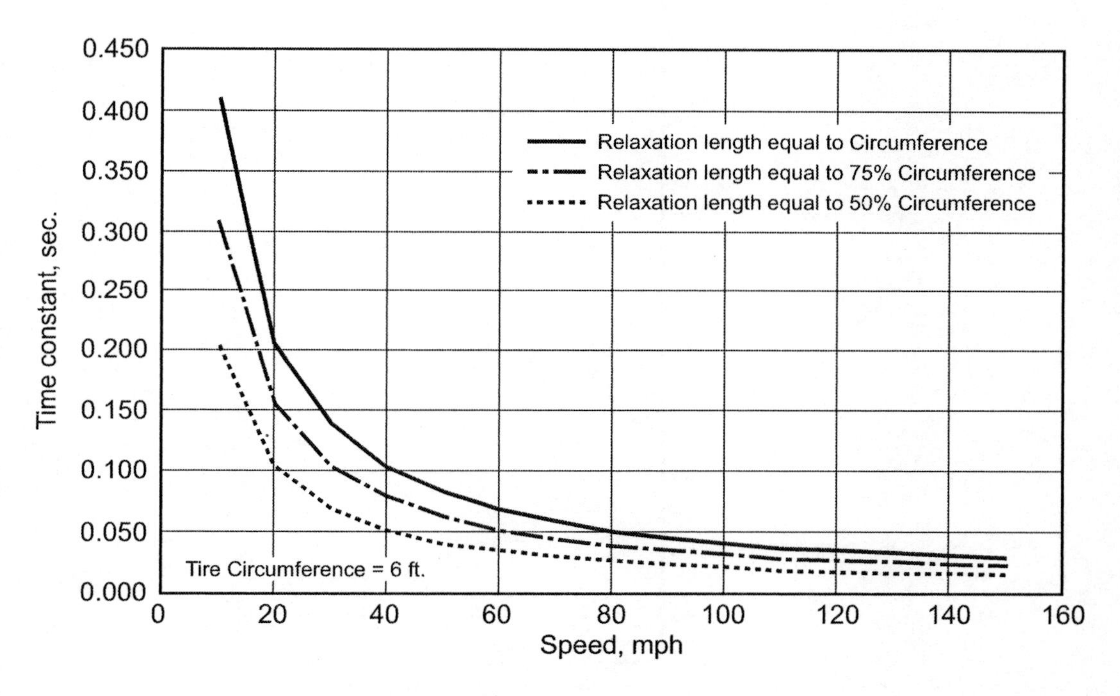

Figure 2.6 *Time constant as a function of speed*

b. The following first-order differential equation can be written:

$$\tau \frac{dF_y}{dt} + F_y = F_{yss} = C_\alpha \alpha$$

where τ is the time constant from a. and F_{yss} is the steady-state value of lateral force. This is a standard first-order equation for which the step response has

[5]Solutions to this question were developed by Steven Radt.

the well-known solution:

$$F_y = C_\alpha \alpha \left(1 - e^{-t/\tau}\right)$$

Note that when $t = \tau$ the equation produces the following result:

$$\frac{F_y}{F_{yss}} = \frac{F_y}{C_\alpha \alpha} = \left(1 - e^{-\tau/\tau}\right) = 1 - \frac{1}{e} \approx 0.632$$

After one time constant the lateral force has risen to 63.2% of its steady-state value. By a similar procedure it can be shown that after four time constants the lateral force is essentially at steady-state. See Figure 2.7.

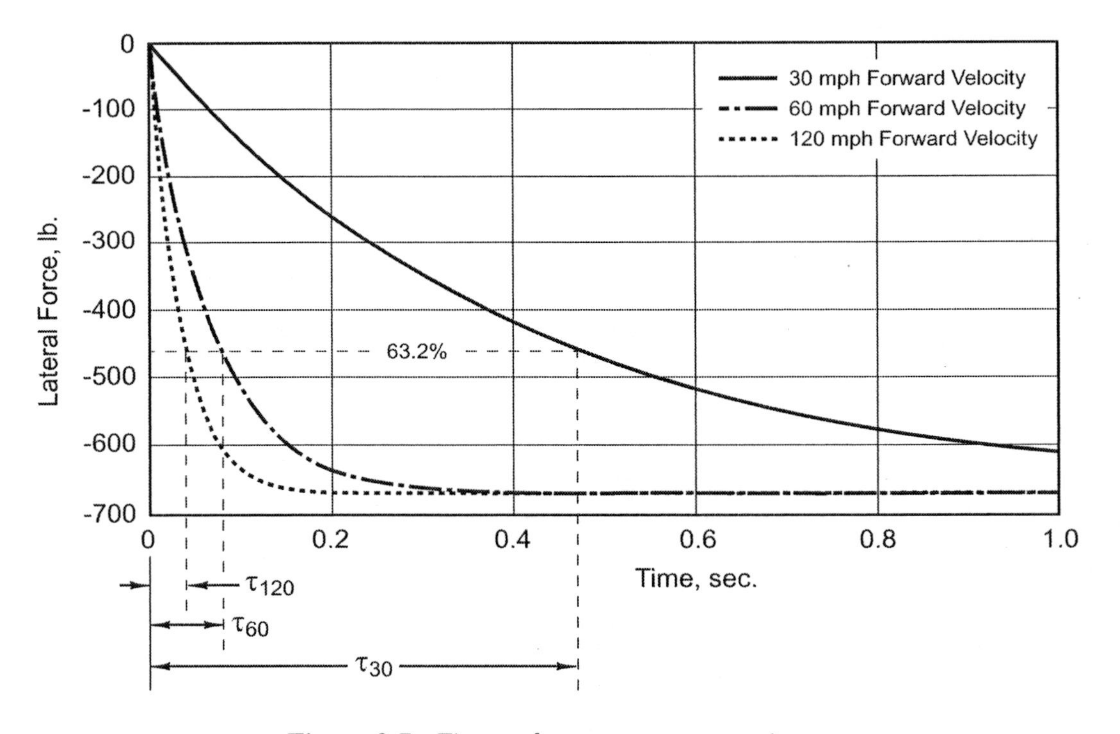

Figure 2.7 *First-order response to step input*

c. Start as in Answer 11b:

$$\tau \frac{dF_y}{dt} + F_y = C_\alpha \alpha(t)$$

and solve this equation using the Laplace transform:

$$\tau\left(s\overline{F_y}(s) - F_y(0)\right) + \overline{F_y}(s) = C_\alpha \overline{\alpha}(s)$$

Let $F_y(0) = 0$ and, for a ramp input, $\overline{\alpha}(s) = a/s^2$ where a is the slope of the ramp.

$$(\tau s + 1)\overline{F_y}(s) = aC_\alpha \Big/ s^2$$

$$\overline{F_y}(s) = \frac{(a/\tau)C_\alpha}{(s + (1/\tau))\,s^2}$$

Evaluating the inverse Laplace transform gives the solution:

$$F_y = a\tau C_\alpha\left(e^{-t/\tau} + t/\tau - 1\right)$$

Without tire lag the lateral force is simply:

$$F_y = C_\alpha \alpha = C_\alpha a t$$

These are compared in Figure 2.8.

In this example the lateral force is assumed to be cornering stiffness times the slip angle. This suffices for small slip angles. A nonlinear model of the lateral force versus slip angle should be used when solving the differential equation for increased accuracy, but this is beyond the scope of the question.

d. Up until now this question has dealt solely with the lag in tire force generation. The tire's time constant is inseparably speed-dependent. At higher speeds the time constant reduces and the time required to reach steady-state lateral force diminishes. The vehicle, however, has its own set of time constants and response lags for various parameters (roll angle, yaw rate, lateral acceleration, etc.). These are typically not as speed-dependent as the tire lag. Furthermore, the vehicle response lags are *always* present.

   At high speeds, the tire lag is much shorter than the vehicle lag so the tire lag may be neglected. At low speeds the time constants for the vehicle parameters and the tire may have similar magnitudes, implying that tire lag can have a significant effect on vehicle response at low speeds. Regardless of the speed, however, there is always a time constant associated with the vehicle motion variables.

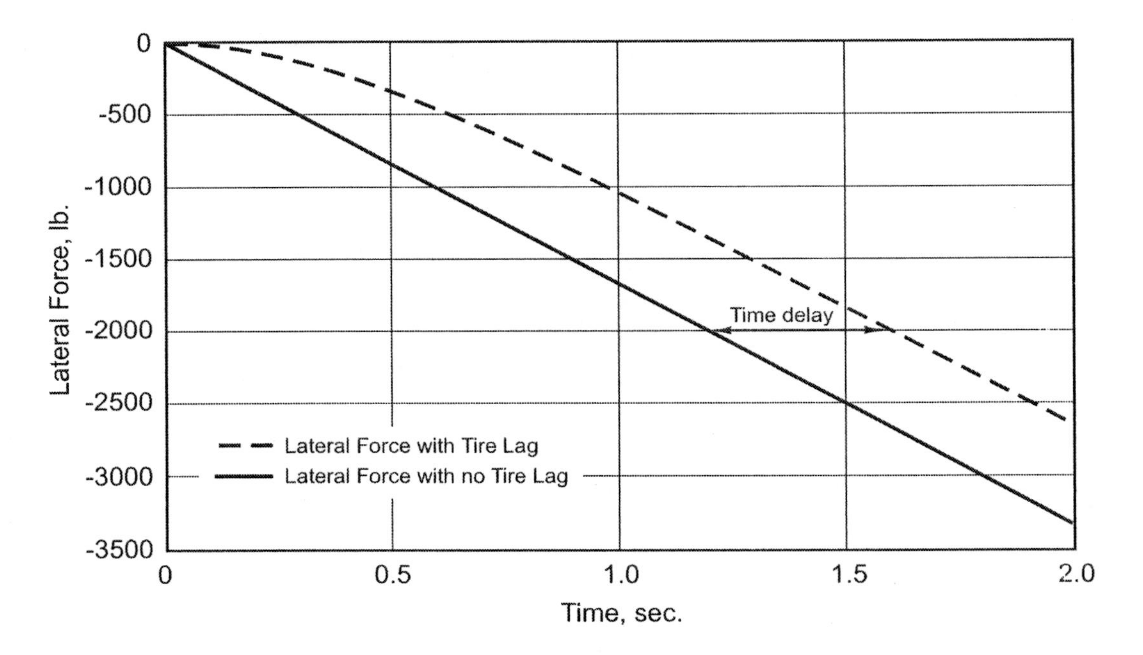

Figure 2.8 *Response to ramp steer with and without lag, 30 mph*

12. a. There are two expressions for lateral force:

$$F_y = k_L \times y \quad \text{and} \quad F_y = C\alpha$$

The first is given in the problem statement. The second is the linear range approximation of lateral force versus slip angle by way of the cornering stiffness. Equating these gives:

$$k_L \times y = C\alpha$$

Slip angle α is the ratio of lateral velocity v to forward velocity u. We can express lateral velocity as the time derivative of lateral displacement. It follows:

$$\alpha = \tan^{-1}\left(\frac{v}{u}\right) \approx \frac{v}{u} = \frac{\dot{y}}{u}$$

Allowing for a steer angle δ:

$$\alpha = \frac{\dot{y}}{u} + \delta$$

Substituting into the original equations leads to the following:

$$k_L \times y = C\alpha$$

$$k_L \times y = C\left(\frac{\dot{y}}{u} + \delta\right)$$

$$\dot{y} - \frac{u \times k_L}{C} y = -\delta u$$

Assuming constant forward speed and steering as an input with respect to time leads to the solution, namely:

$$\dot{y} - \frac{k_L u}{C} y = -u \times \delta(t)$$

b. The equation of interest is a first-order, linear differential equation with constant coefficients. The coefficients k_L and C depend on tire properties and the coefficient u can be considered to be a *parameter* in the solution (we can run the car or tire test machine at any speed we desire). We can also select the time-varying steering input $\delta(t)$ to be anything we desire.

We can solve this differential equation in a myriad of ways. MATLAB provides the following solution for a unit step input:

```
>>pretty(dsolve('Dy-(kl*u*y/C)=u'))

                 kl*u*t
-C/kl  + exp(--------)C1
                   C
```

which is

$$y(t) = -\frac{C}{k_L} + C_1 e^{(k_L ut/C)}$$

where C_1 is a constant based on the initial conditions. We note that since the cornering stiffness is always negative the exponential will decay with time.

When the steer input is a step function the tire lateral force builds exponentially to a final steady-state value with time constant $\tau = -C/k_L u$. After approximately four time constants the lateral force will be practically at steady-state.

c. MATLAB can be used to find the solution for a sinusoidal input of the form $A\sin(\omega t)$ by typing:

```
>>pretty(dsolve('Dy-(kl*u*y/C)=u*A*sin(w*t)'))
```

The result is:

$$y(t) = \frac{-uA\omega C^2\cos(\omega t) - u^2 A k_L C\sin(\omega t)}{k_L^2 u^2 + \omega^2 C^2} + C_1 e^{(k_L ut/C)}$$

where C_1 is a constant based on the initial conditions. Once the transients die out (the exponential terms go to zero) the output is a scaled and time-shifted sine wave with a frequency identical to the input frequency.

d. For a first-order system, we expect the frequency response or Bode plot to exhibit a corner frequency at $1/\tau$ and a final slope of -20 dB/decade. Suppose we choose, from SAE 760030, a combination of tire stiffnesses which results in a $k_L/C = 1.8$ and a forward velocity of 4.7 mph (only a little faster than walking, so that centrifugal effects are negligible). We then get the calculated data shown in Figure 2.9.

When compared with the Weber and Persch data from SAE 760030, the first-order tire model is an excellent predictor of tire dynamic behavior. It also shows that tire dynamics are much, much faster than any expected vehicle dynamics. Even at low speeds the tire time constant is much less than 1 sec.

2.4 Comments on Simple Measurements and Experiments

1. For a typical passenger car tire, changing tire pressure by ±10 psi will result in handling changes noticeable to most drivers. Increasing tire pressure increases the tire's stiffness, reducing the time needed to respond to changes in slip angle. Depending on whether the pressure is increased or decreased on the front and/or rear the vehicle handling will be affected in different ways.

2. At walking speed, the time required for the tire to reach full lateral force at a given slip angle is very large due to the relaxation length. As steering wheel frequency increases, the vehicle tracks a straighter and straighter path. Body roll and vehicle yaw are reduced.

   At 30 mph, the vehicle undergoes large motions (roll, yaw, lateral displacement) with a low frequency steering wheel input. At higher frequencies the vehicle will

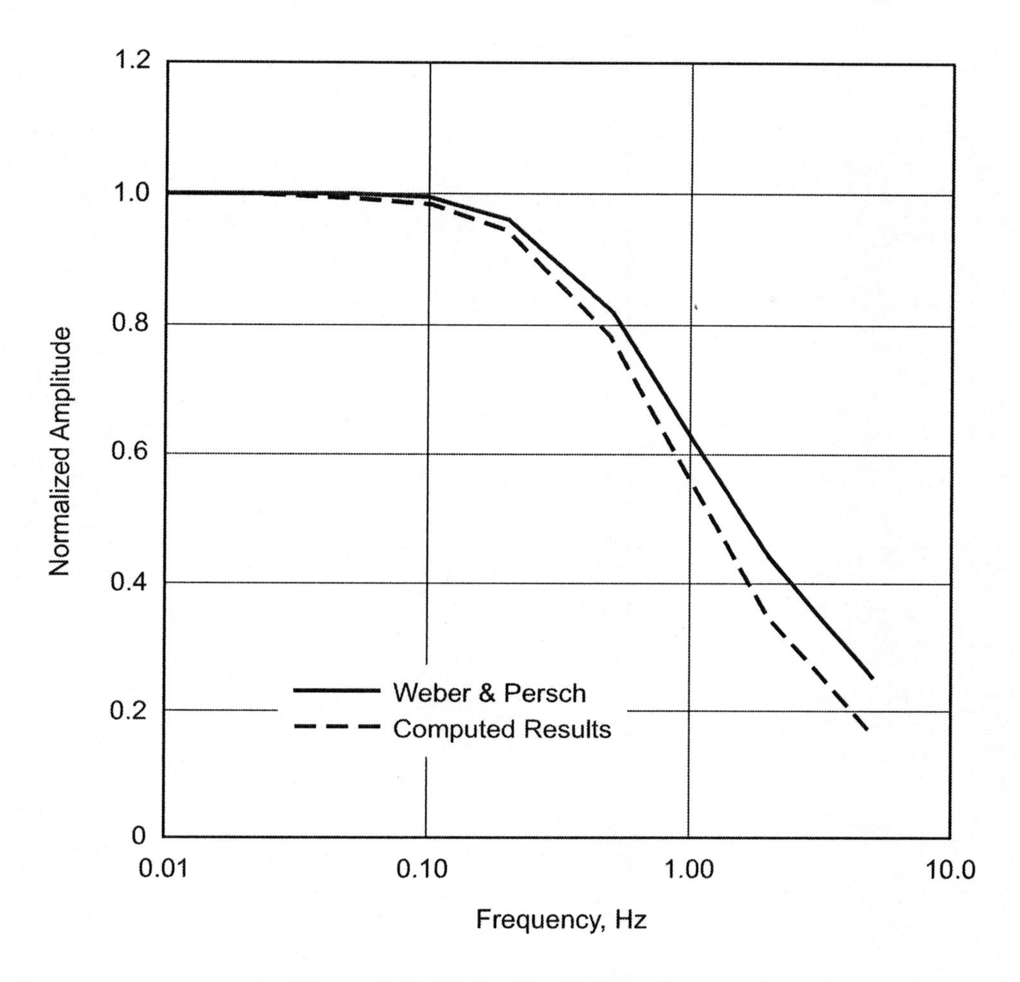

Figure 2.9 *Bode plot of tire response*

follow a straighter path. You may be able to detect the nose moving left and right while the rear follows an essentially straight path. This is because the whole body needs to rotate to produce rear slip angle. High frequency steering doesn't give the front wheels enough time to rotate the whole body for the rear wheels to build up lateral force, so the rear goes straight.

Depending on the car you may also find a resonant steering frequency, or note a phase shift between steering input and response.

3. Free radius is the largest of the three, followed by effective radius and loaded radius. Centrifugal effects increase free radius and loaded radius but (on radial tires) have little effect on effective radius.

There are several ways the print can be measured, depending on the tools available to you. Driving onto a glass plate and viewing the print from below is one technique. Placing the car/tire on pressure sensitive paper would also yield results. With more common means, using a hose to lightly spray all around the base of the tire will leave a dry spot on the pavement which is the shape of the footprint. The same could be done with sawdust or snow. The car may also be jacked up, paint applied to the tire tread and the tire lowered onto the pavement below. Raising the wheel will reveal the footprint.

4. Since aerodynamics become increasingly important as speed increases, performing this test at low speeds can minimize the effect of aerodynamic drag. The effects of road grade can be removed by running the test in two directions over the same section of road and averaging the results. Towing the car has its own difficulties, namely that the force being measured is very small relative to the mass of the car. The towing vehicle will need to provide a very smooth and controllable motive force. A sufficiently stiff spring scale could lead to an oscillation between the tow vehicle and the vehicle being towed.

CHAPTER **3**

Aerodynamic Fundamentals

Solutions to this chapter's problems begin on page 35. Standard temperature and pressure is T = 59°F and p = 1 atm = 14.7 lb./in.^{2}. At these conditions, air has a specific weight γ of 0.0765 lb./ft.^{3} which corresponds to a density ρ of 0.00238 slug/ft.^{3} at standard gravity (32.174 ft./sec.^{2}). See RCVD p.85.

3.1 Problems

1. Use published data (drag coefficient, cross-sectional area) for your own passenger car to create a plot of aerodynamic drag versus forward velocity. Assume air at standard temperature and pressure.

   HINT: To measure cross-sectional area, shine a bright light (headlights will work at dusk) onto the front of the vehicle and trace a shadow outline on an adjacent wall, or take a long-range telephoto photograph of the car seen head-on with an object of known size inserted into the picture.

   If you cannot find data on your car, use the following data for a 1950 Porsche 356A:

   - Drag Coefficient: 0.28
   - Frontal Area: 18.1 ft.^{2}

2. Assume air at standard temperature and pressure is moving past an object with a characteristic length L of 19.7 ft. at 60 mph. Calculate the Reynolds Number for this condition. What airspeed is required to test a 1/4-scale model and have similar flow to the full-scale object?

3. Assume a rolling resistance coefficient of 0.05, a drag coefficient of 0.35, a total vehicle weight of 3000 lb., a cross-sectional area of 18 ft.^{2} and air at standard temperature and pressure. At what forward speed does aerodynamic drag equal rolling resistance for this vehicle?

4. Assume a flat, level roadway with a tire-road friction coefficient of 0.75. Suppose a particular race car generates 500 lb. of downforce at 100 mph. Calculate the increase in speed which this vehicle can achieve on a 100 ft. radius skidpad when compared to the speed achievable without aerodynamics. You will have to extract lift versus velocity information from the above by assuming that $-\text{Downforce} = \text{Lift} = \text{k}V^2$. Repeat for a skidpad radius of 500 ft. Is there a more dramatic improvement at this speed? Explain.

5. For a particular car traveling at 175 mph, pressure taps have been located at the front and rear of the car, and at locations of 1/4, 1/2 and 3/4 of the distance rearward from the front of the car—all on its upper surface. Suppose the car is driven in still air at standard temperature and pressure on a smooth, flat, level roadway and the following measurements of equivalent airflow velocity at each tap location have been obtained:

Pressure Tap Location	Equivalent Velocity
Front	175 mph
1/4 distance	205 mph
1/2 distance	260 mph
3/4 distance	228 mph
Rear	181 mph

   a. Make a plot of pressure coefficient versus longitudinal vehicle position.

   b. Assume the car in question has an effective plan area 4 ft. wide and 8 ft. long, weighs 3000 lb. and has a 50/50 front/rear static weight distribution. The front and rear wheels are located at the front and rear ends of the effective plan area. What are the wheel loads and weight distribution at 175 mph?

6. A given vehicle weighs 3000 lb., has a frontal area of 18 ft.^{2}, a drag coefficient of 0.35 and a rolling resistance coefficient of 0.05. Calculate the total aerodynamic drag force at 75 mph under two different operating conditions:

a. At Long Beach, California, which is at sea level and has an air temperature of 59°F, i.e., under standard conditions.

b. At Pikes Peak International Raceway outside Colorado Springs, Colorado, at an altitude of 5357 ft. and with an air temperature of 82°F.

Solution of (b) will require use of a table of the "Standard Atmosphere"[1].

7. Suppose the hatchback vehicle depicted in RCVD Figure 3.8, p.98, has a cross-sectional area of 17.5 ft.^{2} and is traveling in air at standard temperature and pressure. Calculate the total aerodynamic drag for hatch angles ranging from 10 deg. to 60 deg. in increments of 5 deg., and at forward speeds of 30 mph (representing city driving) and at 60 mph (representing freeway driving). Make a plot of total drag force versus hatch angle for both speeds. Based on the results of this calculation, would you expect a vehicle designer to be able to obtain significant fuel economy improvements through optimization of the hatch angle at city speeds? How about at freeway speeds? How about at racing speeds approaching 150 mph?

8. Aerodynamic downforce is proportional to the square of velocity. Lateral acceleration is also proportional to the square of velocity. It follows that a sufficiently large amount of aerodynamic downforce would allow for any magnitude of acceleration, but this doesn't happen. Why?

3.2 Simple Measurements and Experiments

> ATTENTION: All experiments should be conducted at legal speeds and under safe conditions. A freeway makes a good aerodynamic test track on early Sunday mornings when traffic is light or nonexistent. These experiments can be conducted on a dry road, as tire wear is inconsequential.

1. Using your own passenger car, wash the car thoroughly, dry it completely and wax it. Locate a dusty driving environment and drive in it until the car is completely covered with a fine layer of dust. An alternative way to obtain a dust covering is to fill an old sock with flour or baking powder and dust the entire car with it. Using a hose with a nozzle which can be set on the finest possible spray pattern, lightly

[1] An online source of the Standard Atmosphere table is provided by Public Domain Aeronautical Software at http://www.pdas.com/e2.htm

spray the car, being sure to not allow water to build up so much that it runs off the car. Immediately drive the car at freeway speeds before allowing the water to dry and note the flow field and flow patterns around the various surfaces and features of the car. Compare with RCVD Figure 3.12, p.104.

2. Drive your own passenger car to 60 mph. Depress the clutch or place the automatic transmission in neutral and allow the car to coast down to a very slow speed (be sure traffic conditions behind are safe so that this is not a danger to you or other drivers). Try to pick a smooth and level section of roadway and a calm day to do this experiment. Have an assistant record time and velocity values as you read them from the speedometer, or read them directly into a tape recorder. While an accelerometer may be useful, the deceleration levels associated with this test are quite small. Make a plot of the characteristic coastdown curve for your vehicle. Do the same test in the opposite direction to average out grade irregularities and any steady wind.

3. Make up a 1.0 $ft.^2$ piece of stiff cardboard. With an assistant driving at freeway speed, hold the cardboard outside an open window of your passenger car. Note the differences in drag when the plate is perpendicular to the airflow and when it is parallel to the airflow. Suppose the plate is presented to the airflow at an angle; now what do you observe? Be prepared to lose the cardboard!

4. Repeat the coastdown test of Experiment 2 with your car windows up and then with them down. Repeat again with your trunk lid propped completely open. Can you measure and/or observe differences in the coastdown curve for your particular passenger car in these different configurations? Would you expect to? Consider making other temporary modifications and rerunning the test, such as:

   a. Taping over all window, hood and door cracks.

   b. Covering wheel wells by taping cardboard over them.

   c. Temporarily covering the radiator opening on the front of the car.

   d. Removing the radio aerial, mirrors and other protuberances.

3.3 Problem Answers

1. Calculate the drag force as a function of velocity:

$$\text{Drag} = \frac{1}{2}\rho C_D A V^2 = \frac{1}{2}\left(0.00238\ \text{slug/ft.}^3\right)(0.28)\left(18.1\ \text{ft.}^2\right)V^2$$

$$\text{Drag} = \left(6.031 \times 10^{-3}\right)V^2 \text{ with V in ft./sec.}$$

$$\text{Drag} = \left(2.803 \times 10^{-3}\right)V^2 \text{ with V in mph}$$

Plotting at speeds from 0 to 140 mph results in Figure 3.1.

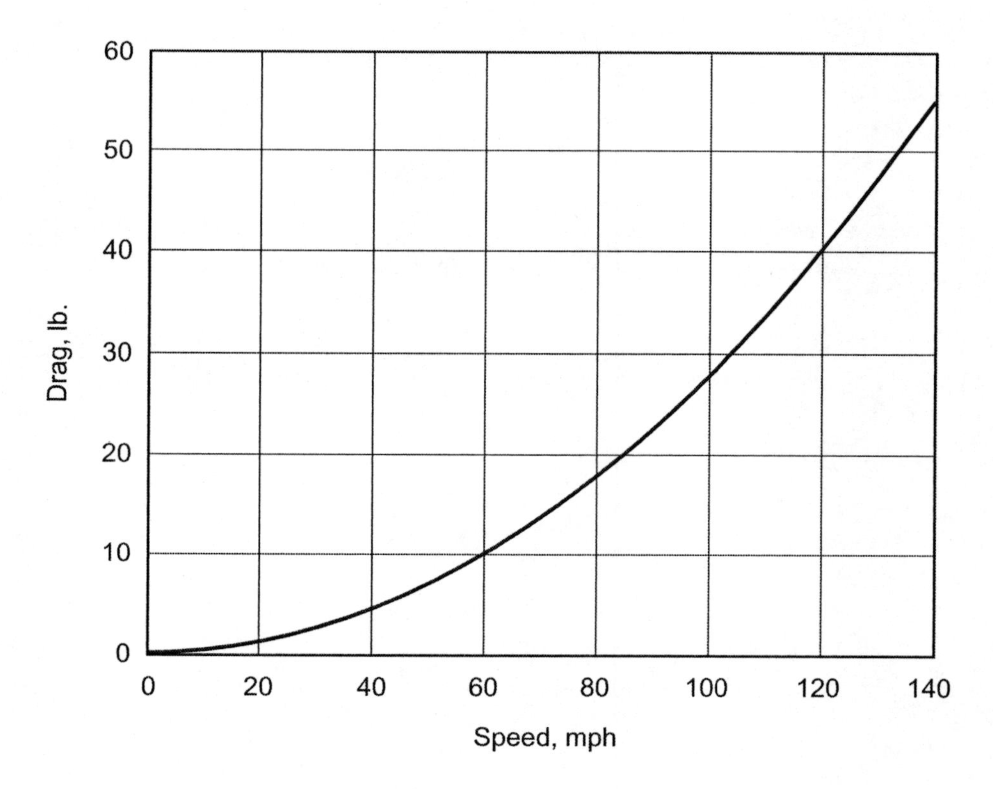

Figure 3.1 *Drag curve for a 1950 Porsche 356A*

2. For the full-size object:

$$Re = \frac{\rho VL}{\mu} = \frac{VL}{\nu} = \frac{(88 \text{ ft./sec.})(19.7 \text{ ft.})}{0.00016 \text{ ft.}^2\text{/sec.}} = 1.0835 \times 10^7$$

For flow similarity, the Reynolds Number of the 1/4 scale model must be identical to this figure.

$$1.0835 \times 10^7 = \frac{V(4.925 \text{ ft.})}{0.00016 \text{ ft.}^2\text{/sec.}}$$

$$V = 352.4 \text{ ft./sec.} = 240 \text{ mph}$$

3. Assume rolling resistance is independent of speed and equate rolling resistance with aerodynamic drag.

$$\text{Rolling Resistance} = \text{Aerodynamic Drag}$$

$$C_{RR} \times W = \frac{1}{2}\rho C_D A V^2$$

$$0.05 \times 3000 \text{ lb.} = \frac{1}{2}\left(0.00238 \text{ slug/ft.}^3\right)(0.35)\left(18 \text{ ft.}^2\right)V^2$$

$$V = 141.5 \text{ ft.}/\text{sec.} = 96.5 \text{ mph}$$

4. First, solve for k:

$$\text{Downforce} = kV^2$$

$$-500 \text{ lb} = k(100 \text{ mph})^2 \Rightarrow k = -0.05$$

Speed on the skidpad is found from two equations for lateral force:

$$F_y = \mu F_z \quad \text{where} \quad F_z = W - \text{Lift} = W - kV^2$$

$$F_y = ma_y = \frac{W}{g}\frac{V^2}{R}$$

Combining these and solving for V gives:

$$V = \sqrt{\frac{\mu W}{(W/gR) + \mu k}}$$

Using this equation leads to the following results:

Radius ft.	k –	Speed ft./sec.	Speed mph
100	0	49.1	33.5
100	−0.05	50.7	34.5
500	0	109.8	74.9
500	−0.05	131.4	89.6

The aerodynamic downforce provides a ~3% increase in speed on the 100 ft. radius and a ~20% increase on the 500 ft. radius. Downforce is proportional to the square of velocity, so on the larger radius where the speed is higher, downforce provides a larger increase in normal load, resulting in a correspondingly larger increase in speed.

5. a. Apply the equation at the bottom of RCVD p.107 to the tabular data to determine the pressure coefficient.

$$C_p = 1 - \left(\frac{V}{V_\infty}\right)^2$$

Location	C_p
0.00	0.000
0.25	−0.372
0.50	−1.207
0.75	−0.697
1.00	−0.070

These data are plotted in Figure 3.2.

b. To answer this part, first assume the pressure coefficient under the bottom of the car is constant at 1.0. This is unrealistic, as flow beneath the car in ground effect will change the underbody pressure coefficient. A more complete solution would include the pressure change due to underbody flow.

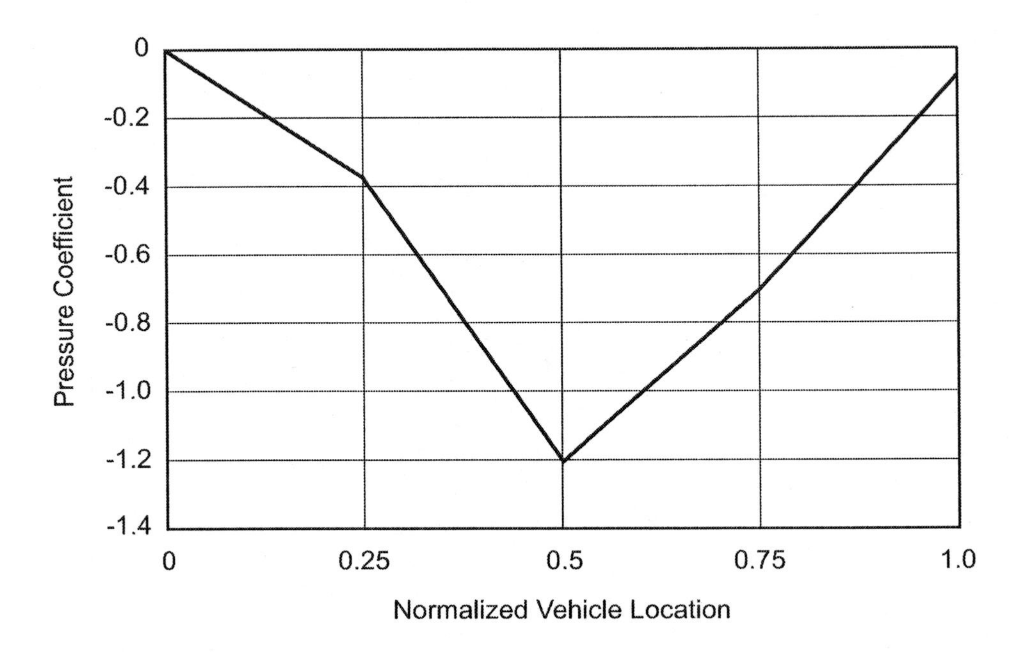

Figure 3.2 *Pressure coefficient along vehicle length*

Upper surface pressure coefficients greater than the underbody coefficient produce downforce, while those less than the underbody coefficient produce lift. In this case, the pressure taps indicate lift.

$$C_p = \frac{\Delta p}{q_\infty} \text{ where } q_\infty = \frac{1}{2}\rho V_\infty^2$$

Assume the pressures are constant over the areas as shown in Figure 3.3.

Location ft.	Δp lb./ft.^{2}
0.0	0.000
2.0	−29.2
4.0	−94.6
6.0	−54.6
8.0	−5.5

The total force is:

$$F = pA = (0.0)4 + (-29.2)8 + (-94.6)8 + (-54.6)8 + (-5.5)4 = -1449 \text{ lb.}$$

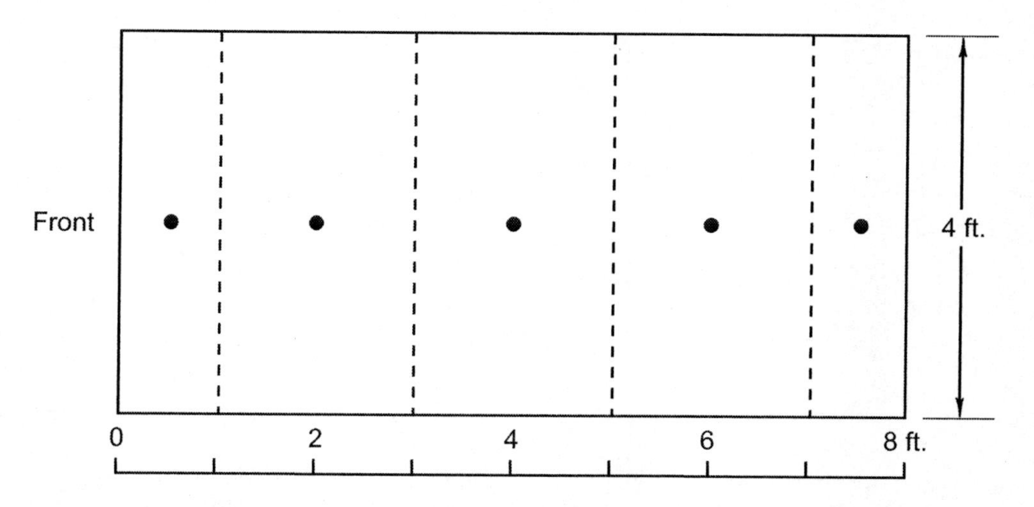

Figure 3.3 *Division of plan area into regions with centroids*

Note that this calculation is actually the discretization of an integral taken over the upper surface of the vehicle. The center of pressure is determined by weighting the force on each section at that section's centroid:

$$C_p = \frac{\Sigma x_i F_i}{\Sigma F_i} = \frac{0.5\,(0) + 2\,(-233.6) + 4\,(-756.8) + 6\,(-436.8) + 7.5\,(-22)}{-1449.2}$$

which results in $C_p = 4.33$ ft. This is 0.33 ft. behind the CG, so the rear wheels will experience more lift than the front. Taking moments about the front of the effective plan area gives the rear wheel loads:

$$(4\text{ ft.})\,(3000\text{ lb.}) = (4.33\text{ ft.})\,(1449.2\text{ lb.}) + (8\text{ ft.})\,W_R \Rightarrow W_R = 715.6\text{ lb.}$$

The sum of the front and rear wheel loads equals the static wheel load plus the aerodynamic downforce:

$$W_F + 715.6\text{ lb.} = 3000\text{ lb.} - 1449.2\text{ lb.} \quad \Rightarrow \quad W_F = 835.2\text{ lb.}$$

At 175 mph, the wheel loads are about half the static wheel loads and the Front/Rear weight distribution is approximately 54/46.

6. a. Since these operating conditions match standard conditions, use the standard value for the density of air.

$$\text{Drag} = \frac{1}{2}\rho C_D A V^2$$

$$\text{Drag} = \frac{1}{2}\left(0.002378 \text{ slug/ft.}^3\right)(0.35)\left(18 \text{ ft.}^2\right)\left(75 \text{ mph} \times \frac{88 \text{ ft./sec.}}{60 \text{ mph}}\right)^2$$

$$\text{Drag} = 90.6 \text{ lb.}$$

b. Here, the operating conditions do not match with standard conditions, so the air density needs to be calculated. At an altitude of 5357 ft., the Standard Atmosphere table indicates a density 85.24% of that at sea level and 59°F. The standard temperature at this altitude is 40°F. The standard atmosphere at 5357 ft. and 40°F gives:

$$\rho_{5357,40} = (0.8524)\,\rho_{0,59} = (0.8524)(0.002378) = 0.002027 \text{ slug/ft.}^3$$

The question says the air temperature is 82°F. The density needs to be corrected for this temperature. This can be done using the formula for temperature change on RCVD p.86.

$$\rho_{5357,40} = \left(\frac{T_{40}}{T_{82}}\right)\rho_{5357,82} = \left(\frac{40+460}{82+460}\right)0.002027 = 0.001870 \text{ slug/ft.}^3$$

Finally, use this new density to calculate the drag:

$$\text{Drag} = \frac{1}{2}\left(0.001870 \text{ slug/ft.}^3\right)(0.35)\left(18 \text{ ft.}^2\right)\left(75 \text{ mph} \times \frac{88 \text{ ft./sec.}}{60 \text{ mph}}\right)^2$$

$$\text{Drag} = 71.3 \text{ lb.}$$

7. Use the formula $\text{Drag} = 0.5\rho A C_d V^2$ with a density of 0.00238 slug/ft.^{3}, an area of 17.5 ft.^{2} and speed measured in feet per second. Drag coefficients are read off RCVD Figure 3.8 for various hatch angles. Considering speeds of 30, 60 and 150 mph (44, 88 and 220 ft./sec., respectively) produces the results shown in the following table and in Figure 3.4.

Hatch Angle deg.	C_d –	Drag 30 mph lb.	Drag 60 mph lb.	Drag 150 mph lb.
10	0.330	13.3	53.2	332.4
15	0.335	13.5	54.0	337.4
20	0.350	14.1	56.4	352.5
25	0.400	16.1	64.5	402.9
30	0.430	17.3	69.3	433.1
35	0.395	15.9	63.7	397.8
40	0.395	15.9	63.7	397.8
50	0.395	15.9	63.7	397.8
60	0.395	15.9	63.7	397.8

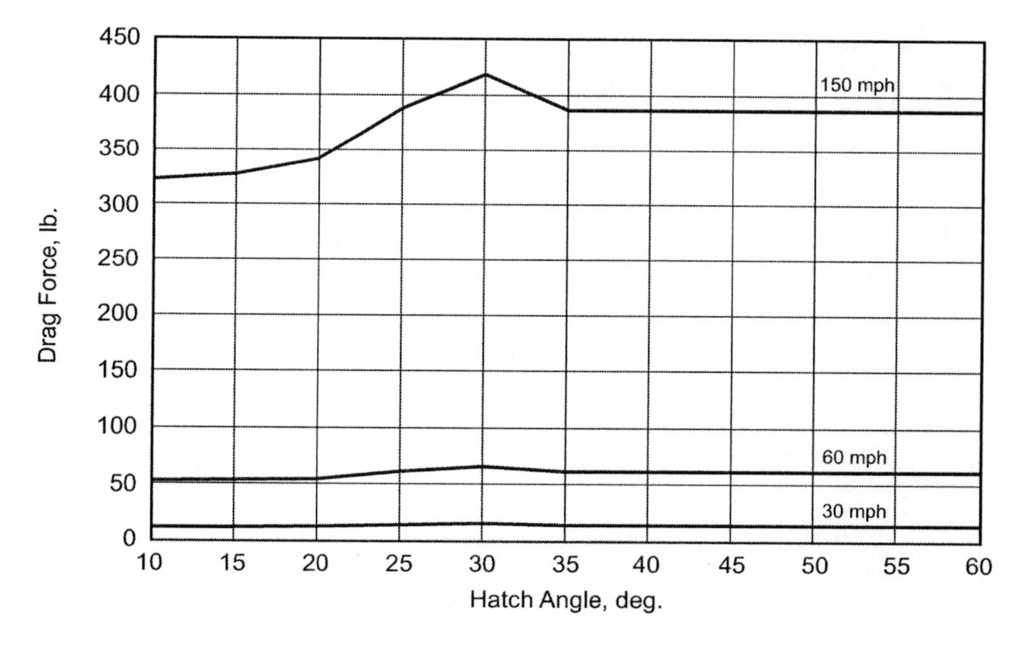

Figure 3.4 *Effect of hatch angle on drag*

At city speeds, the hatch angle has a negligible effect on drag and, thus, fuel economy. At freeway speeds, hatch angle has a larger effect than at city speeds, but it is still fairly small. At racing speeds, the effect of hatch angle on drag is dramatic.

8. Downforce increases in proportion to the velocity squared. Lateral acceleration is proportional to lateral force which, on a constant radius, must increase in proportion to velocity squared. The problem statement's reasoning is based on a

hidden (and false) assumption, namely that lateral force increases linearly with downforce. Tire load sensitivity precludes this behavior. As load increases the friction coefficient decreases, so increases in downforce result in decreasingly large increases in lateral force.

The result is this: As speed increases, downforce increases in proportion to velocity squared, but the lateral force does not keep pace with the downforce. Lateral acceleration is therefore not unlimited.

3.4 Comments on Simple Measurements and Experiments

1. Techniques similar to this are used in wind tunnels where colored dyes or pressure sensitive paints can be introduced to visualize the flow field. These same techniques are sometimes used during on-track testing.

2. Note that this test inherently includes the effect of rolling resistance. For an interesting discussion and a method for extracting your car's drag coefficient from this test see SAE Paper No. 720099 by White and Korst, *The Determination of Vehicle Drag Contributions from Coastdown Tests.*

3. Many children have performed this experiment by placing their hand out the window of a moving car. The use of cardboard increases the surface area, thereby increasing the forces. Parallel to the airflow the drag force is relatively small. Perpendicular to the airflow, the drag is very large. At an angle, a significant amount of lift is generated along with the drag force.

4. While all these changes have an effect, you may not be able to notice all of them. As with most aerodynamics, the effects are more noticeable at higher speeds. For example, some CART and IRL teams routinely tape over the engine cover and sidepod joints to help reduce drag when they race on superspeedways.

CHAPTER **4**

Vehicle Axis Systems

This chapter deals mostly with vehicle dynamics standards and conventions developed by the automotive industry and published by SAE, for example, *Vehicle Dynamics Terminology*[1] and *Vehicle Aerodynamics Terminology*[2].

Solutions to this chapter's problems begin on page 45.

4.1 Problems

1. Make isometric sketches of the SAE Vehicle Axis System, the SAE Tire Axis System and the SAE Aerodynamic Axis System. Suggest reasons underlying the choice of origin and orientation for each of these axis systems.

2. The SAE axis system is not the only axis system in use. Can you name any others? Discuss the pros and cons of these systems relative to the SAE axis system.

3. A vehicle on a flat road is in a steady right hand turn. The body has a roll angle of -2 deg. and no pitch angle. The left front tire's road contact point is located 4 ft.

[1] *Vehicle Dynamics Terminology*, SAE J670e, July 1976.
[2] *Vehicle Aerodynamics Terminology*, SAE J1594, June 1987.

ahead, 2.5 feet to the left and 1.5 ft. below the vehicle CG. The left front is steered +5 deg. cambered −3 deg. and has the following forces measured in the Tire Axis System: $F_X = 200$ lb., $F_Y = 400$ lb. and $F_Z = -800$ lb.

Determine the equivalent forces and moments at the CG in the Vehicle Axis System. (Hint: This problem is best solved via the "Euler Angle Transformation Matrix".)

4. Consider a modestly understeering vehicle cornering at steady-state on a constant radius right hand turn. For each of the following, state the sign (positive or negative) for the given quantity:

   a. Longitudinal Velocity, u
   b. Lateral Velocity, v
   c. Yaw Velocity, r
   d. Longitudinal Acceleration, A_X
   e. Lateral Acceleration, A_Y
   f. Tire Longitudinal Forces, F_X
   g. Tire Lateral Forces, F_Y
   h. Vehicle Yaw Moment, M_Z
   i. Steer Angle, δ
   j. Tire Slip Angles, α
   k. Vehicle Sideslip Angle, β
   l. Body Roll Angle, ϕ

5. For a stability analysis, is it necessary to consider the Earth Fixed Axis System?

4.2 Simple Measurements and Experiments

1. [3]It is common to describe the orientation of an object with respect to a coordinate system through three angles: yaw, pitch and roll. The sequence of these is important, as changing the sequence results in different orientations. The standard order is **Yaw then Pitch then Roll**.

   Conduct the following experiment: Start with this book sitting in front of you on a flat surface (such as a table) such that the text is right-side up (in the normal reading position). Yaw the book +90 deg. The book is now lying flat on the table with its right edge facing you. Now pitch the book +90 deg. This stands the book on end so that someone sitting to your left would be able to read it. Now roll the book +90 deg. The final result is that the book is perpendicular to the table with the originally face-up text visible to you. This sequence is shown in Figure 4.1.

[3]This experiment was inspired by Dr. William Rae.

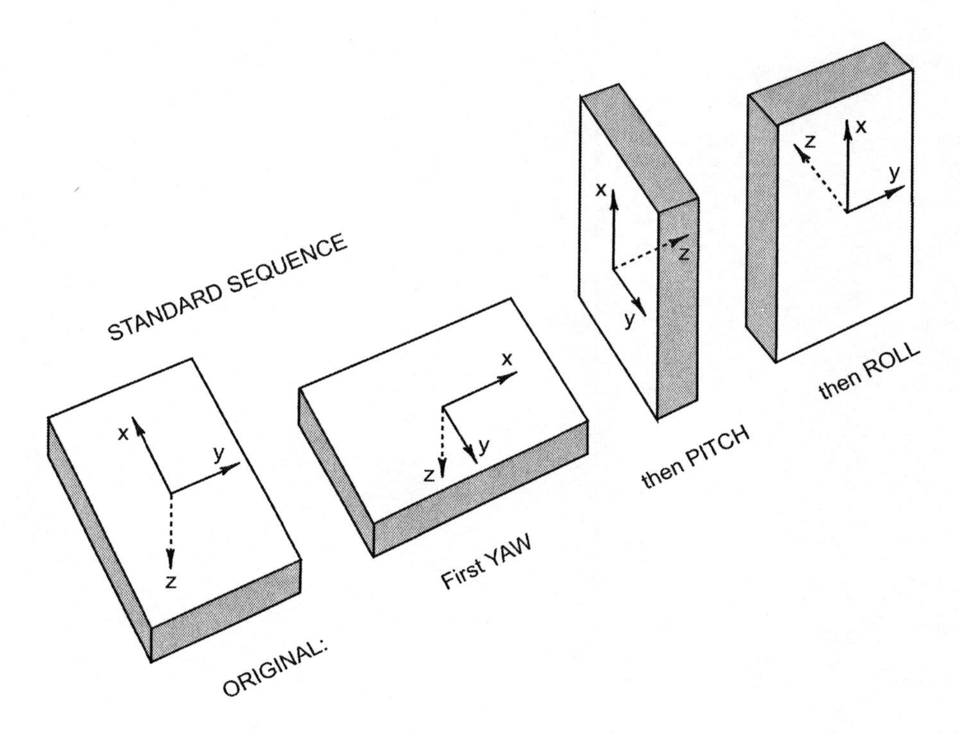

Figure 4.1 Standard sequence of rotations

Repeat the three rotations in different sequences. Convince yourself that different sequences result in different orientations.

4.3 Problem Answers

1. Axis systems are shown in Figure 4.2.

   SAE Vehicle Axis System—The origin is located at the CG where an applied force produces no moment. Forces applied at locations other than the CG can be transferred to the CG by including a corresponding moment. Writing equations about the CG greatly simplifies the vehicle equations of motion. X is positive forward, Y is positive to the right and Z is positive downward such that gravity is a positive force. It is the same as the aircraft stability axis system.

   SAE Tire Axis System—Mimics the Vehicle Axis System, but is centered at a point where the wheel plane intersects the ground plane beneath the tire. X and Y lie in the ground plane and Z is normal to the ground and positive downward. X points

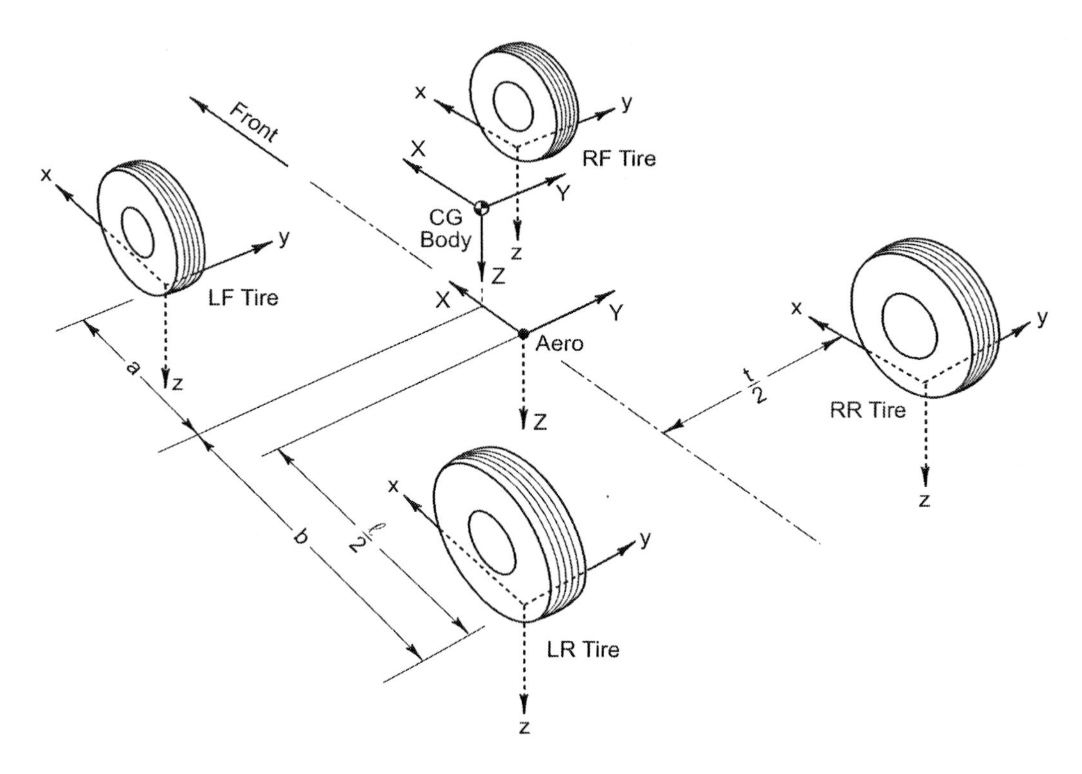

Figure 4.2 *Isometric sketch of SAE axis systems*

forward in the direction of the wheel plane projected into the ground plane. This axis system is convenient because tire forces and moments are described and measured relative to the ground plane and tire orientation.

SAE Aerodynamic Axis System—Aero data is often generated in the wind tunnel using a model, instead of an actual running vehicle. Since the CG of a model will probably not correspond to an actual vehicle (and the CG of the actual vehicle may be unknown), it is convenient to choose the origin at the center of the wheelbase and track. Furthermore, automotive wind tunnels often have the balance (measurement) system under the floor so choosing ground level for the height of the origin is convenient. As with the SAE Vehicle Axis System, positive X is forward along the vehicle centerline, Y is positive to the right and Z is positive downward.

2. There are several non-SAE axis systems in use. The German axis system has been accepted by ISO (as has the SAE system). Here, X is positive forward, Y is positive to the *left* and Z is positive *upward*. The advantage is that Z is positive upward, the way height is normally measured from the ground. Of course, this system also results in gravity being a negative quantity.

In computer graphics (e.g., for driving simulators) it has become common to choose X horizontal across the bottom of the video screen (positive to the right), Y vertical (positive upward) and Z into the screen (so forward motion is positive Z). Note that this is a *left handed* system, where X and Y correspond to the 1st quadrant of the Cartesian (screen) plane. Many mathematical concepts, such as the cross product of two vectors, rely on a right handed coordinate system to produce the correct results. Beware.

Instead of placing the origin at the CG (an "invisible" point), some vehicle applications would benefit from an axis system whose origin was located, say, at the left rear wheel. Distances to other physical parameters would become much easier to accurately measure. Such an axis system would have little application in the aircraft industry as aircraft designers control the CG location very carefully, even before the first prototype airplane is built. In this case the origin can be placed directly at the CG. An axis system centered on the left rear wheel would require transformations to the CG before any calculations are performed.

Whatever vehicle axis system is chosen, it only makes sense for the tire and aerodynamic axis systems to mimic it.

3. First, transform the tire forces from the Tire Axis System to the Vehicle (Body) Axis System. The Euler Angle Transformation Matrix A(ψ, θ and ϕ) is defined as:

$$A = \begin{bmatrix} \cos\theta\cos\psi & \sin\phi\sin\theta\cos\psi - \cos\phi\sin\psi & \cos\phi\sin\theta\cos\psi + \sin\phi\sin\psi \\ \cos\theta\sin\psi & \sin\phi\sin\theta\sin\psi + \cos\phi\cos\psi & \cos\phi\sin\theta\sin\psi - \sin\phi\cos\psi \\ -\sin\theta & \sin\phi\cos\theta & \cos\phi\cos\theta \end{bmatrix}$$

When a point (or vector) is known in an axis system (denoted "1") which is rotated through the angles ψ, θ and ϕ **in that order** relative to another axis system (denoted "2"), the vector can be represented in Axis System "2" as:

$$\begin{Bmatrix} x \\ y \\ z \end{Bmatrix}_2 = A \begin{Bmatrix} x \\ y \\ z \end{Bmatrix}_1$$

The body is rolled −2 deg. from its initial position and the tire is steered +5 deg. from its initial position. With the initial position given the subscript "0":

$$\begin{Bmatrix} x \\ y \\ z \end{Bmatrix}_0 = A_{tire} \begin{Bmatrix} x \\ y \\ z \end{Bmatrix}_{tire} \quad \text{and} \quad \begin{Bmatrix} x \\ y \\ z \end{Bmatrix}_0 = A_{body} \begin{Bmatrix} x \\ y \\ z \end{Bmatrix}_{body}$$

Solving for the body coordinates in terms of the tire coordinates:

$$\begin{Bmatrix} x \\ y \\ z \end{Bmatrix}_{body} = A'_{body} A_{tire} \begin{Bmatrix} x \\ y \\ z \end{Bmatrix}_{tire}$$

where A′ denotes the inverse of matrix A.

$$\begin{Bmatrix} x \\ y \\ z \end{Bmatrix}_{body} = A'(0,0,-2)\,A(5,0,0) \begin{Bmatrix} x \\ y \\ z \end{Bmatrix}_{tire}$$

Applying this transformation to the tire forces gives:

$$\begin{Bmatrix} F_x \\ F_y \\ F_z \end{Bmatrix}_{body} = A'(0,0,-2)\,A(5,0,0) \begin{Bmatrix} F_x \\ F_y \\ F_z \end{Bmatrix}_{tire}$$

$$\begin{Bmatrix} F_x \\ F_y \\ F_z \end{Bmatrix}_{body} = \begin{bmatrix} 1 & 0 & 0 \\ 0 & 0.9994 & -0.0349 \\ 0 & 0.0349 & 0.9994 \end{bmatrix} \begin{bmatrix} 0.9962 & -0.0872 & 0 \\ 0.0872 & 0.9962 & 0 \\ 0 & 0 & 1 \end{bmatrix} \begin{Bmatrix} 200 \\ 400 \\ -800 \end{Bmatrix}_{tire}$$

$$\begin{Bmatrix} F_x \\ F_y \\ F_z \end{Bmatrix}_{body} = \begin{Bmatrix} 164.4 \\ 443.6 \\ -785.0 \end{Bmatrix}$$

These are the tire forces expressed in the body axis system centered at the left front tire contact point. Transferring these to the CG will entail the calculation of moments as the forces are translated. The required moment arms are:

$$\begin{Bmatrix} x \\ y \\ z \end{Bmatrix} = A'(0,0,-2) \begin{Bmatrix} 4 \\ -2.5 \\ 1.5 \end{Bmatrix} = \begin{Bmatrix} 4 \\ -2.551 \\ 1.4118 \end{Bmatrix}$$

The vector cross product of the forces with the moment arms gives the required moments about the CG:

$$\begin{Bmatrix} M_x \\ M_y \\ M_z \end{Bmatrix} = \begin{Bmatrix} F_x \\ F_y \\ F_z \end{Bmatrix} \times \begin{Bmatrix} -x \\ -y \\ -z \end{Bmatrix} = \begin{Bmatrix} 1376.3 \\ 3372.1 \\ 2193.8 \end{Bmatrix} \text{ lb.-ft.}$$

4. a. Longitudinal Velocity—When the vehicle is traveling forward, u is positive.

   b. Lateral Velocity—Positive

   c. Yaw Velocity—Positive

   d. Longitudinal Acceleration—Steady-state cornering on a constant radius implies constant forward speed, so A_X is zero. If the vehicle were increasing its speed A_X would be positive.

   e. Lateral Acceleration—Positive

   f. Tire Longitudinal Forces—To maintain constant speed in the presence of induced drag there must be some tractive effort applied by the tires, so F_x is positive.

   g. Tire Lateral Forces—Positive

   h. Vehicle Yaw Moment—Steady-state cornering on a constant radius implies constant yaw velocity, so M_Z must be zero.

   i. Steer Angle—Positive

   j. Tire Slip Angles—Negative

   k. Vehicle Sideslip Angle—The sign depends on whether or not the vehicle is above or below the tangent speed as explained on RCVD p.174. Below the tangent speed β is positive.

   l. Body Roll Angle—Negative

5. The Earth Fixed Axis System is needed only when calculating vehicle location and path—as in a driving simulator. Otherwise, it is sufficient from a stability analysis or other vehicle dynamics standpoint to look at the vehicle response in terms of lateral acceleration, yaw rate, etc.

4.4 Comments on Simple Measurements and Experiments

1. While any angle would work, +90 deg. is perhaps the easiest to visualize. Figure 4.1 showed the standard sequence: Yaw then Pitch then Roll. Compare against a different sequence, say, pitch then yaw then roll as shown in Figure 4.3.

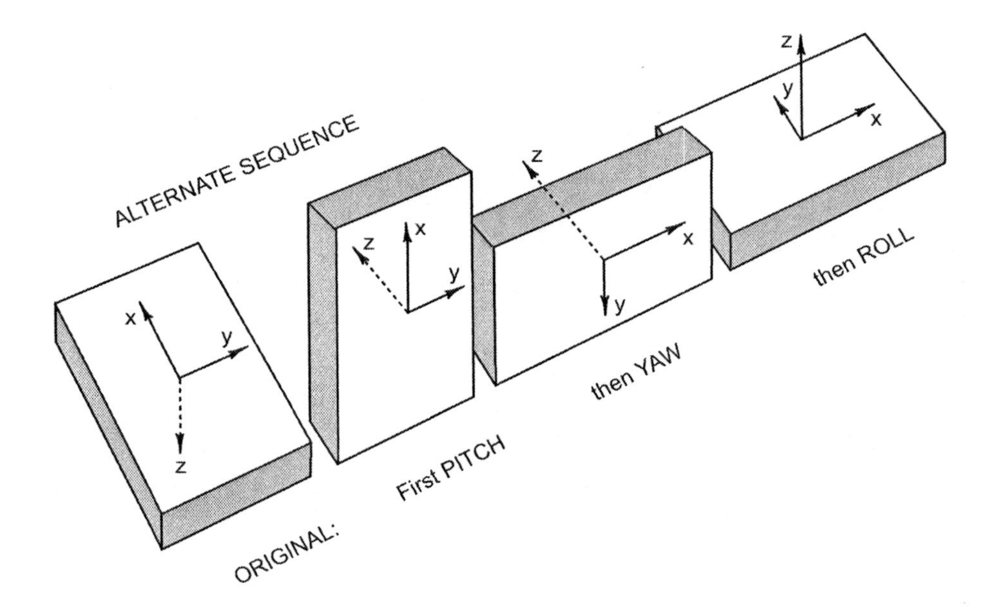

Figure 4.3 *Non-standard sequence resulting in different final position*

The result is a different orientation from that shown in Figure 4.1.

CHAPTER **5**

Simplified Steady-State Stability and Control

Solutions to this chapter's problems begin on page 55.

Unless otherwise stated, the problems in this chapter use the following bicycle model vehicle data. This vehicle is referred to as the "**CH5**":

- Wheelbase = 104.5 in.
- Total weight = 3700 lb.
- 57/43 Front/Rear weight distribution
- Front cornering stiffness, $C_F = -640$ lb./deg.
- Rear cornering stiffness, $C_R = -580$ lb./deg.
- CG height of the vehicle = 20 in. above ground

5.1 Problems

1. For the CH5, calculate the Ackermann steer angle required to travel a 100 ft. diameter circle. Suppose a steering gear has a 14:1 ratio (i.e., 14 deg. at the steering

wheel produces 1 deg. at the front wheel). What is the Ackermann steer angle at the steering wheel? Does this angle depend on forward speed? Is the steering wheel angle equal to the slip angle of the tire? Why (not)?

2. [1]RCVD Figure 5.5, p.131, is drawn for a low forward velocity (below the tangent speed) and assumes road load power (constant speed). Redraw the figure for a speed above tangent speed.

3. The CH5 is traveling at 30 mph along a straight road which, due to the crown, is tilted 5 deg. in this lane (the right side of the car is lowered relative to a flat road). Determine the steering wheel angle and vehicle sideslip angle required to maintain a path parallel to the road's centerline.

4. The CH5 is traveling on a straight, flat road. A steady crosswind applies a 100 lb. force 1.5 ft. in front of the CG. Calculate the steady-state steer angle needed to counteract this crosswind such that the car travels in a straight line.

5. What is the highest speed the CH5 can obtain while navigating a 100 ft. radius skidpad at steady-state? To answer this question, assume the following tire characteristics, where C_α is the front or rear cornering stiffness, as needed:

$$F_y = \begin{cases} C_\alpha \alpha & \text{when} \quad -5 < \alpha < 5 \text{ deg.} \\ C_\alpha (5 \text{ deg.}) & \text{when} \quad \alpha > 5 \text{ deg.} \\ C_\alpha (-5 \text{ deg.}) & \text{when} \quad \alpha < -5 \text{ deg.} \end{cases}$$

Figure 5.1 interprets this equation graphically.

This equation for lateral force introduces a maximum output to the linear tire model, approximating the peak of a real tire's cornering force curve. The cornering stiffnesses in this tire model remain unchanged from the CH5 data at the beginning of the chapter.

6. a. Suppose you could move the CG of the CH5. Where would you put it to make the CH5 neutral steer? Assume the tire cornering stiffnesses are load-independent.

   b. Now suppose the cornering stiffness of a certain tire is given by the equation:

$$C_\alpha = -0.4F_z + \left(0.8 \times 10^{-4}\right) F_z^2$$

---

[1]This problem suggested by Frank Winchell.

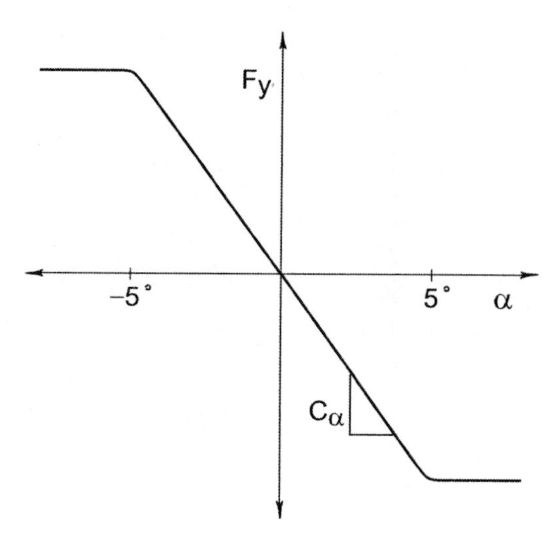

Figure 5.1 *Breakaway slip angle to introduce a peak lateral force*

Suppose this tire was placed on both ends of the CH5. If you wanted to make the CH5 neutral steer with these tires, where would you put the CG?

7. a. Consider the CH5 traveling at 30 mph. Calculate the stability derivatives Y_β, Y_r, Y_δ, N_β, N_r and N_δ. Does this vehicle understeer or oversteer in steady-state? Explain.

   b. Suppose the weight distribution is reversed such that it is 43/57 front/rear. Recalculate the stability derivatives and compare with Question 7a.

8. A certain vehicle is defined by $a = 6$ ft., $b = 4$ ft., $C_F = C_R = -350$ lb./deg. Specify new values of a, b, C_F and C_R such that the static directional stability of the new vehicle has the same magnitude but opposite sign as the original vehicle while preserving the yaw damping. (That is, make $N_{\beta 2} = -N_{\beta 1}$ and $N_{r2} = N_{r1}$, where "1" is the original vehicle and "2" is the new vehicle.)

9. Determine the stability factor K for the CH5. Calculate this car's characteristic speed or critical speed.

10. Calculate the steady-state yaw rate response to control (r/δ) for the CH5 at speeds from 0 to 150 mph. Make a plot similar to RCVD Figure 5.20, p.163.

11. For the CH5 calculate the neutral steer point. Calculate the static margin. Does the static margin indicate understeer or oversteer? Make a sketch of the vehicle showing the wheels, center of gravity and neutral steer point.

12. The linear equations of motion in derivative notation for two degrees of freedom, namely yaw rate (r) and vehicle sideslip (β), are given by RCVD Equation 5.10, p.150. These equations correspond to a vehicle with non-rolling sprung mass. Write a new set of equations in derivative notation which include the roll degree of freedom (ϕ). Identify the "spring", "damper" and "coupling" terms of the roll degree of freedom.

13. The concepts of "over/understeer" and "yaw damping" are frequently confused or misunderstood. What derivative, in the linear range, best describes each? Describe each in physical terms and cite a mechanical analogy for each.

5.2 Simple Measurements and Experiments

NOTE: All experiments should be carried out at low speeds (say, below 30 mph) under safe conditions. A shopping market or campus parking lot on early Sunday morning often makes a good skidpad.

WARNING: The following experiments can produce significant amounts of tire wear. Do not continue the experiments for many laps unless you are unconcerned about tire wear. Tire wear can be reduced by performing the experiments on wet or snow covered surfaces.

You may want to make several preliminary measurements and calculations before beginning, such as Ackermann Steer Angle, wheelbase measurement and others which are used in the bicycle model of the automobile. See RCVD p.127.

1. Consider your personal vehicle. From measurements or your Owner's Manual determine the front/rear static weight distribution of the vehicle. Determine the bicycle model parameters a and b. Determine the recommended tire pressures and observe whether or not the tires at all four wheels are the same size and mounted on the same-sized rims. Based on these observations, does your vehicle understeer or oversteer? What important factors does the bicycle model omit?

2. Lay out a 50 ft. radius circle on a large, flat surface with chalk and/or traffic cones. Drive around the circle at constant velocity for several forward velocities and have a colleague time your laps. Consider speeds of 10, 15 and 20 mph and compare your speedometer reading with the speed calculated from the lap time. Calculate the lateral accelerations achieved and compare these with your

own driving experiences. Does 20 mph seem to produce a high lateral acceleration? Note that most drivers never intentionally use more than about 0.3 to 0.4g lateral acceleration during normal driving!

3. Using tape, chalk, etc., mark an angular reference frame on your steering wheel so that a back seat passenger can observe steering wheel angle. Drive on a 50 ft. circle at 10, 15 and 20 mph, measuring the steering wheel angle at each speed. Make a plot similar to that of RCVD Figure 5.20, p.163. Does your car understeer or oversteer? How can you tell?

4. Repeat Experiments 2 and 3 above with two different operating conditions:

   a. Front tires 10 psi over recommended tire pressure, rear tires 10 psi under recommended tire pressure.

   b. Rear tires 10 psi over recommended tire pressure, front tires 10 psi under recommended tire pressure.

5. On a wet or snowy surface, drive your vehicle around in a circle at a safe speed which is significantly below its cornering capability under the conditions chosen. Maintain a constant steer angle while slowly and continuously applying power. If your vehicle is front wheel drive, what do you notice? If it is rear wheel drive, what do you notice?

5.3 Problem Answers

1. Using the small angle assumption, the Ackermann steer angle is calculated as:

$$\delta_{\text{Ack}} = \frac{\ell}{R} = \frac{104.5/12}{50} = 0.174 \text{ rad.}$$

which is 9.98 deg. at the front wheel. Multiplying by the steering ratio (14:1) gives the steering wheel angle, 139.7 deg.

The Ackermann steer angle is a geometric construct which does not depend on forward speed. Slip angles, however, *do* depend on forward speed. Note the velocity dependence of the middle term in the slip angle equation (front):

$$\alpha_F = \beta + \frac{ar}{V} - \delta$$

2. Figure 5.2 is drawn for cornering above the tangent speed. Compared with RCVD Figure 5.5, p.131, the angle of the velocity vector has changed, reversing the sign of β. Also, the path of the rear tires now lies outside the path of the front tires.

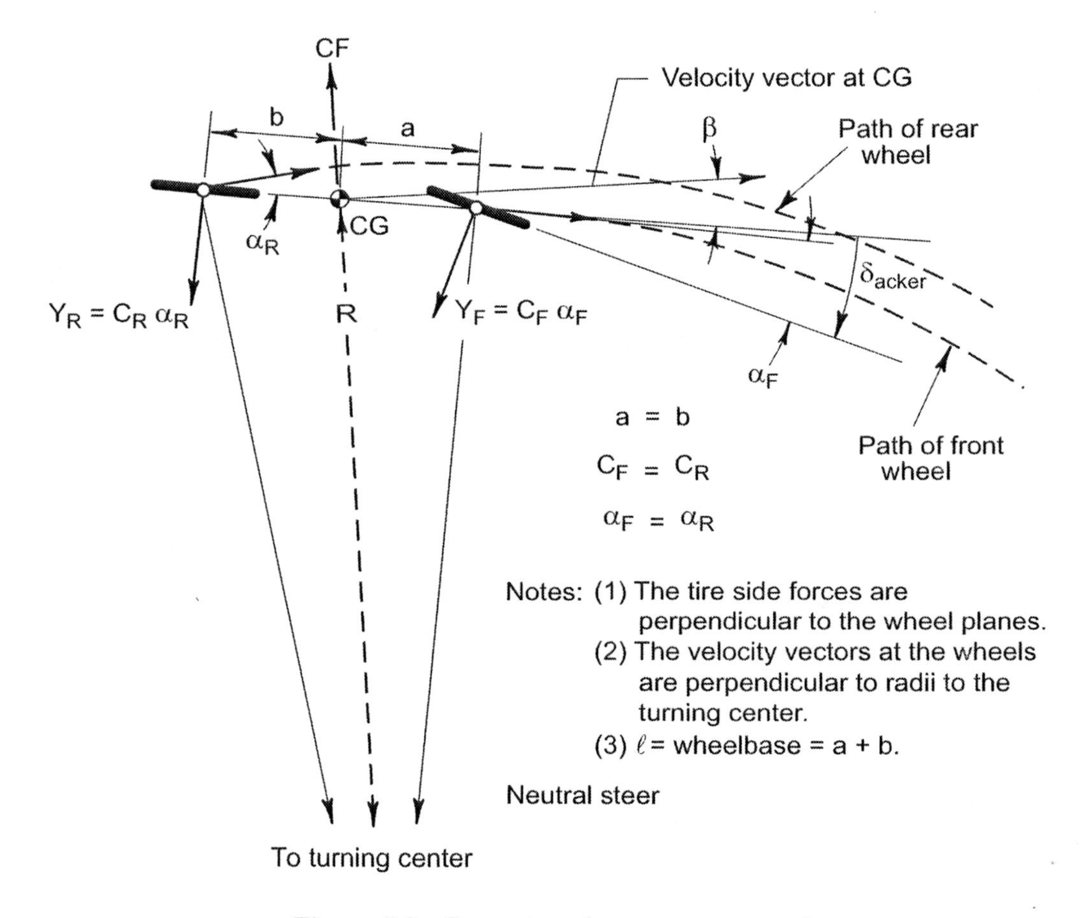

Figure 5.2 *Cornering above tangent speed*

3. The tires need to provide a total lateral force equal to $-3700 \sin(5 \text{ deg.})$ or -322.5 lb. The quantity is negative since the force is to the left. This force is divided between the front and rear tires according to the weight distribution:

$$F_{yF} = F_y\left(\frac{b}{\ell}\right) = -322.5\,(0.57) = -183.8 \text{ lb.}$$

$$F_{yR} = F_y\left(\frac{a}{\ell}\right) = -322.5\,(0.43) = -138.7 \text{ lb.}$$

Back-calculate the slip angles from the lateral forces and the cornering stiffness at each end of the vehicle:

$$F_{yF} = C_{\alpha F}\alpha_F \quad \Rightarrow \quad \alpha_F = \frac{-183.8}{-640} = 0.287 \text{ deg.}$$

$$F_{yR} = C_{\alpha R}\alpha_R \quad \Rightarrow \quad \alpha_R = \frac{-138.7}{-580} = 0.239 \text{ deg.}$$

Apply RCVD Equations 5.3 and 5.4, pp.147-148. Since the vehicle is traveling straight ahead, set $r = 0$:

$$\alpha_R = \beta - \frac{br}{V} \quad \Rightarrow \quad \beta = \alpha_R = 0.239 \text{ deg.}$$

$$\alpha_F = \beta + \frac{ar}{V} - \delta \quad \Rightarrow \quad \delta = \beta - \alpha_F = 0.239 - 0.287 = -0.048 \text{ deg.}$$

The vehicle's nose is pointed toward the road centerline and the front wheels are steered slightly toward the road centerline. Note that this calculation is independent of vehicle speed since the path is straight.

4. Start by translating the wind force to the CG, resulting in a force and a moment $F_y = 100$ lb. and $M_z = 150$ lb.-ft. The tire forces must equilibrate with these forces and moments:

$$F_{yF} + F_{yR} = -100 \text{ lb.}$$

$$aF_{yF} - bF_{yR} = -150 \text{ lb.- ft.}$$

These equations are solved simultaneously: $F_{yF} = -74.3$ lb. and $F_{yR} = -25.7$ lb. Calculating slip angles for these lateral forces based on the cornering stiffnesses:

$$F_{yF} = C_{\alpha F}\alpha_F \quad \Rightarrow \quad \alpha_F = \frac{-74.3}{-640} = 0.116 \text{ deg.}$$

$$F_{yR} = C_{\alpha R}\alpha_R \quad \Rightarrow \quad \alpha_R = \frac{-25.7}{-580} = 0.044 \text{ deg.}$$

For a straight path the rear slip angle is equivalent to the vehicle sideslip angle, so $\beta = 0.044$ deg. At the front:

$$\delta = \beta - \alpha_F = -0.072 \text{ deg.}$$

The entire vehicle is pointed slightly into the crosswind and the front tires are steered into the wind.

5. Begin by calculating the maximum lateral force that can be produced by the front and rear tires. This occurs at a slip angle of ± 5 deg.

$$F_{yF\,max} = C_{\alpha F}\alpha_F = -640(-5) = 3200 \text{ lb.}$$

$$F_{yR\,max} = C_{\alpha R}\alpha_R = -580(-5) = 2900 \text{ lb.}$$

Note that the question did not specify which direction the car was turning. Since the bicycle model is symmetric it doesn't matter—only the signs of the solutions will change. The solution below is presented for a right hand turn. Thus, lateral forces are positive.

Calculate a yaw balance with both the front and rear tires saturated:

$$N = aF_{yF} - bF_{yR} = (3.74)(3200) - (4.96)(2900) = -2416 \text{ lb.-ft.}$$

But, at steady-state $N = 0$. The above result indicates that the moment from the rear has a larger magnitude than the moment from the front. Thus, the front of the vehicle is limiting and the CH5 has terminal plow. Since the front is already at its maximum lateral force, the lateral force on the rear needs to be *reduced* to restore the yaw moment balance:

$$0 = aF_{yF} - bF_{yR} = (3.74)(3200) - (4.96)F_{yR} \quad \Rightarrow \quad F_{yR} = 2413 \text{ lb.}$$

This puts the rear slip angle at $2413/-580 = -4.16$ deg.

To calculate the speed, sum the wheel forces and divide by the vehicle weight to determine the lateral acceleration in g-units:

$$A_Y = \frac{F_{yF} + F_{yR}}{W} = \frac{3200 + 2413}{3700} = 1.51 \text{ g}$$

which is 48.8 ft./sec.^{2}. Since the radius is known the speed is:

$$A_y = \frac{V^2}{R} \quad \Rightarrow \quad V = \sqrt{48.8 \times 100} = 69.8 \text{ ft./sec.} = 47.6 \text{ mph}$$

6. a. For neutral steer, $N_\beta = 0$. Thus:

$$aC_F = bC_R = (\ell - a)\,C_R$$

$$a = \frac{\ell C_R}{C_F + C_R} = \frac{(8.7)(-580)}{-640 - 580} = 4.14 \text{ ft.}$$

For neutral steer, a = 4.14 ft., b = 4.57 ft.

b. Again, for neutral steer, $N_\beta = 0$. Thus:

$$aC_F = bC_R$$

$$a\left[-0.4\left(\frac{Wb}{\ell}\right) + \left(0.8\times 10^{-4}\right)\left(\frac{Wb}{\ell}\right)^2\right] = b\left[-0.4\left(\frac{Wa}{\ell}\right) + \left(0.8\times 10^{-4}\right)\left(\frac{Wa}{\ell}\right)^2\right]$$

After much algebra this leads to the simple conclusion:

$$a = b$$

So, a = 8.7/2 = 4.35 ft. and b = 4.35 ft.

7. a. First, determine a and b.

$$a = 0.43 \times (104.5/12) = 3.74 \text{ ft.}$$

$$b = 0.57 \times (104.5/12) = 4.96 \text{ ft.}$$

Use these and V = 30 mph = 44 ft./sec. to calculate the stability derivatives:

$$Y_\beta = C_F + C_R = -640 + -580 = -1220 \text{ lb./deg.}$$

$$Y_r = \frac{1}{V}\left(aC_F - bC_R\right)$$

$$= \frac{1}{44}\left(3.74(-640) - 4.96(-580)\right) = 10.98 \text{ lb./deg./sec.}$$

$$Y_\delta = -C_F = 640 \text{ lb./deg.}$$

$$N_\beta = aC_F - bC_R = 3.74(-640) - 4.96(-580) = 483.2 \text{ lb.-ft./deg.}$$

Note that $N_\beta > 0$, so the vehicle exhibits steady-state understeer.

$$N_r = \frac{1}{V}\left(a^2 C_F + b^2 C_R\right)$$

$$= \frac{1}{44}\left(3.74^2(-640) + 4.96^2(-580)\right) = -527.7 \text{ lb.-ft./deg./sec.}$$

$$N_\delta = -aC_F = -3.74(-640) = 2393.6 \text{ lb.-ft./deg.}$$

b. Reversing the weight distribution results in a = 4.96 ft. and b = 3.74 ft. New values are calculated in the table below and compared with the previous values:

Stability Derivatives	Previous Values Part 7a	New Values Part 7b	Units
Y_β	−1220	−1220	lb./deg.
Y_r	+10.98	−22.80	lb./deg./sec.
Y_δ	+640	+640	lb./deg.
N_β	+483.2	−1005.2	lb.-ft./deg.
N_r	−527.7	−542.2	lb.-ft./deg./sec.
N_δ	+2393.6	+3174.4	lb.-ft./deg.

Note that the vehicle of Part 7b has $N_\beta < 0$, indicating oversteer.

8. Indicate values for the original vehicle with subscript “1” and the new vehicle with subscript “2”. Noting that $C_{F1} = C_{R1} = C_1$:

$$N_{\beta 1} = a_1 C_{F1} - b_1 C_{R1} = C_1 (a_1 - b_1)$$

$$N_{r1} = \frac{1}{V}\left(a_1^2 C_{F1} + b_1^2 C_{R1}\right) = \frac{C_1}{V}\left(a_1^2 + b_1^2\right)$$

Any change in C_1 above will alter the magnitude of the expressions. We don’t want to change the magnitude of either expression. Thus, the new cornering stiffnesses should be the same as the old cornering stiffnesses, $C_2 = C_1$.

For the new vehicle:

$$N_{\beta 2} = -N_{\beta 1} = -C_1\,(a_1 - b_1) = C_1\,(b_1 - a_1)$$

But $N_{\beta 2} = C_1\,(a_2 - b_2)$, so by direct comparison $a_2 = b_1$ and $b_2 = a_1$. Check to see that the yaw damping is preserved:

$$N_{r2} = \frac{C_1}{V}\left(a_2^2 + b_2^2\right) = \frac{C_1}{V}\left(b_1^2 + a_1^2\right) = N_{r1}$$

Yaw damping is preserved. Substituting the original values into these relationships gives the new values: $a_2 = 4$ ft., $b_2 = 6$ ft. and $C_{F2} = C_{R2} = -350$ lb./deg.

9. The wheelbase is $\ell = 104.5/12 = 8.7$ ft. To determine the mass:

$$m = \frac{W}{g} = \frac{3700 \text{ lb.}}{32.174 \text{ ft./sec.}^2} = 115 \text{ slugs}$$

Use the stability derivatives calculated in Problem 7a in the calculation of K. Convert the values of the stability derivatives into units per radian (from units per degree). Applying the formula for the stability factor from RCVD p.161 gives:

$$K = \left(\frac{mN_\beta}{\ell\left(N_\beta Y_\delta - Y_\beta N_\delta\right)}\right)$$

$$= \left(\frac{115\,(27686)}{8.7\,((27686)\,(36669) - (-69901)\,(137134))}\right) = 3.45 \times 10^{-5} \text{ sec.}^2/\text{ft.}^2$$

Since $K > 0$ the CH5 is an understeering vehicle. Understeering vehicles do not have a critical speed. The characteristic speed is:

$$V_{char} = \sqrt{\frac{1}{K}} = \sqrt{\frac{1}{3.45 \times 10^{-5}}} = 170.2 \text{ ft./sec.} = 116 \text{ mph}$$

10. In terms of stability derivatives, RCVD Equation 5.15, p.155, gives:

$$\frac{r}{\delta} = \frac{Y_\beta N_\delta - N_\beta Y_\delta}{N_\beta Y_r - N_\beta mV - Y_\beta N_r}$$

Alternatively, RCVD Equation 5.43, p.161, gives the yaw rate response to control in terms of the stability factor, K:

$$\frac{r}{\delta} = \frac{V/\ell}{1+KV^2}$$

Problem 9 calculated K as 3.45×10^{-5} sec.^{2}/ft.^{2} and the CH5 has a wheelbase of 8.7 ft. Using these values results in Figure 5.3.

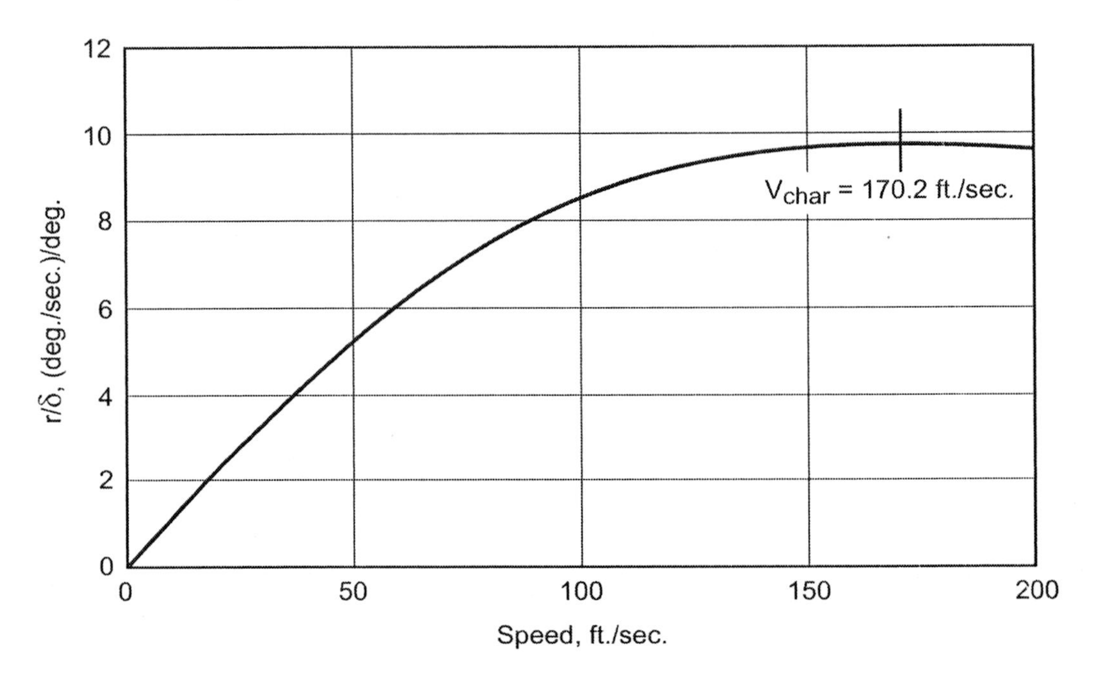

Figure 5.3 *CH5 yaw rate response to control*

11. For the bicycle model, the neutral steer point is the ratio of the rear cornering stiffness to the total cornering stiffness.

$$NSP = \frac{C_R}{C_F + C_R} = \frac{580}{640+580} = 0.475$$

Static margin compares the neutral steer point to the normalized CG location:

$$SM = NSP - \frac{a}{\ell} = 0.475 - 0.430 = 0.045$$

Since the static margin is positive it indicates understeer.

In terms of the wheelbase, the CG is located 44.9 in. aft of the front wheel and the neutral steer point is located 49.6 in. aft of the front wheels. This is sketched in Figure 5.4.

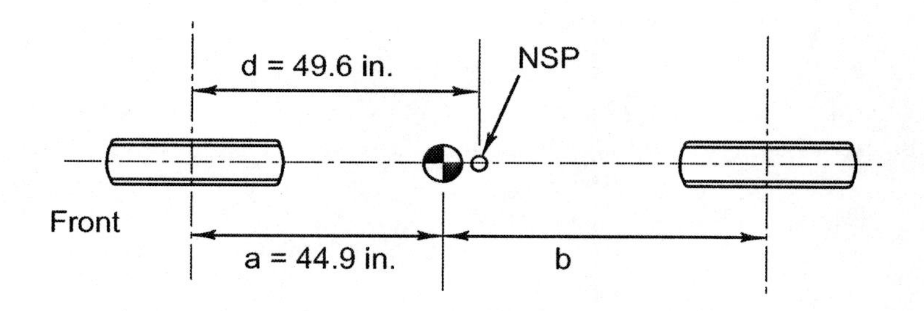

Figure 5.4 *Neutral steer point in relation to CG*

12. The existing lateral force and yawing moment equations now contain additional terms to account for the roll angle:

$$mV\left(r+\dot{\beta}\right)+m_s h\ddot{\phi}=Y_\beta\beta+Y_r r+Y_\phi\phi+Y_\delta\delta$$

$$I_z\dot{r}+I_{xz}\ddot{\phi}=N_\beta\beta+N_r r+N_\phi\phi+N_\delta\delta$$

Here, m_s is the sprung mass, h is the distance from the sprung CG to the roll axis, I_{xz} is the yaw-roll product of inertia, $\ddot{\phi}$ is roll acceleration, $\dot{\phi}$ is roll velocity and ϕ is the roll angle.

A third equation is needed for the roll degree of freedom[2]:

$$I_x\ddot{\phi}+I_{xz}\dot{r}+m_s hV\left(r+\dot{\beta}\right)=L_\beta\beta+L_r r+L_\phi\phi+L_\delta\delta$$

The terms on the right hand side can be tailored to the needs of the model. For example, while the L_ϕ term is always important for the automobile (it's the spring term), coupling terms such as L_r and L_δ are typically insignificant and therefore are commonly omitted. Another coupling term, L_β, is not very important for the automobile (except at high speeds), but is a significant derivative in airplane equations of motion. Furthermore, it is common to add the term $L_{\dot{\phi}}\dot{\phi}$ to the roll equation—this is a roll-damping term.

[2]The bicycle model with roll degree of freedom was originally developed by Leonard Segel in *Theoretical Prediction and Experimental Substantiation of the Response of the Automobile to Steering Control*, Equations 35-37 on p.31, Institution of Mechanical Engineers, London, England, 1958.

For example, if the model is for calculating dynamic response, the roll degree of freedom could be expressed as:

$$I_x\ddot{\phi} + I_{xz}\dot{r} + m_s hV\left(r + \dot{\beta}\right) = L_\beta \beta + L_\phi \phi + L_{\dot{\phi}}\dot{\phi}$$

13. Oversteer/understeer is best described by the derivative N_β. This is the yawing moment due to vehicle sideslip. It is analogous to a weathervane. Yaw damping is best described by the derivative N_r. This is the yawing moment due to yawing velocity. It is analogous to a rotational damper.

5.4 Comments on Simple Measurements and Experiments

1. The bicycle model is a simple, linearized model of the automobile and its tires. From this model a large body of terminology is derived. This terminology is applied to more advanced models or to the actual vehicle (e.g., understeer, yaw damping, etc.).

   Most passenger cars have the same tires on all four corners. Without measured tire data for your particular vehicle you cannot know the magnitude of the tire load sensitivity effects on cornering stiffness. If you assume equal cornering stiffnesses and note that the CG of your vehicle is ahead of the mid-wheelbase point you can use the concepts of the bicycle model to predict that your vehicle understeers (passenger cars typically understeer).

   A more accurate model (especially for predicting US/OS) would include roll stiffness distribution as discussed in later chapters of RCVD. The accuracy of any vehicle model is also heavily dependent on the quality of the tire model. As the complexity of the vehicle model increases, however, the ease at which phenomenon can be explained and understood decreases. The bicycle model, while far short of a comprehensive vehicle model, is useful in that it allows the fundamentals of vehicle behavior and vehicle modeling to be studied in the simplest meaningful environment.

2. Timing the laps is important—speedometers do not accurately measure speed along the path while cornering. If the centerline of the car is following a radius of 50 ft. the following lateral accelerations result:

Speed (mph)	10	15	20
Acceleration (g)	0.13	0.31	0.53

3. Working through the dimensional analysis gives yawing velocity, r = 84V/R, where r is in deg./sec., V is in mph and R is in ft. If the data points in your plot fall on a straight line your car is neutral steer. If the trend is concave upward your car is oversteer. If the trend is concave downward your car is understeer.

   You may find it interesting to repeat this experiment after placing a heavy load in the trunk.

4. In general, increasing tire pressure increases cornering stiffness while decreasing tire pressure decreases cornering stiffness. Thus, the car in 4a is likely to be less understeer (more oversteer) than the baseline car. The car in 4b is likely to be more understeer (less oversteer) than the baseline car.

5. Refer to RCVD Figures 5.55 to 5.57, p.220 to 223, for a hint about what to expect. Path curvature at constant steer angle changes with driving thrust. The rear wheel drive vehicle of RCVD Figure 5.5 spins-out at high steer angles coupled with high driving thrust.

CHAPTER 6

Simplified Transient Stability and Control

Solutions to this chapter's problems begin on page 71.

6.1 Problems

1. Derive the equation of motion for a spring-mass-damper (SMD) system which is forced in two different ways:

   a. *Force Excitation*: Motion is created by applying a force directly to the mass which is attached to one end of the spring. The other end of the spring is fixed to ground or inertial space. See RCVD Figure 6.3, p.238, for an example of this method of forcing the system.

   b. *Base Excitation*: Motion is created by applying a displacement to the end of the spring opposite to the end at which the mass is attached. This model, a single wheel forced by a road displacement (input), is often termed a quarter-car ride model.

   c. What is the order of each model (first, second, ...)?

d. Do the transfer functions of each model have the same denominator or characteristic equation?

e. Do the transfer functions of each model have the same numerator?

2. Suppose a two degree-of-freedom vehicle has the following characteristic equation in yaw:

$$s^2 + 1.4s + 13 = 0$$

a. Calculate the undamped natural frequency, damped natural frequency and damping ratio for this vehicle.

b. Is the vehicle stable or unstable?

c. Is the vehicle underdamped, overdamped, critically damped or undamped?

d. Would you expect overshoot of the vehicle yaw response when the input to the vehicle is a step steering command by the driver?

e. Plot the location of the poles of this characteristic equation in the s-plane.

3. Consider the vehicle of the "First Example" on RCVD p.256:

- Weight, $W = 2000$ lb.
- Wheelbase, $\ell = 8.33$ ft.
- Yaw inertia, $I_Z = 832$ slug-ft.^{2}
- 65/35 Front/Rear weight distribution
- Front cornering stiffness, $C_F = -240$ lb./deg.
- Rear cornering stiffness, $C_R = -200$ lb./deg.

For this vehicle, calculate the transfer functions $\overline{T}_{\delta,u}$ and $\overline{T}_{\delta,r}$ at a speed of 40 mph. Then, draw a Bode plot of amplitude ratio (or gain) in dB and phase versus $\log_{10}\omega$ for normalized yaw rate r/δ.

4. A human operator can sinusoidally steer at 1-2 Hz with an amplitude of perhaps ±45 deg. Review your results for Problem 3. Based on those results, can a human operator ever force the vehicle of Problem 3 at resonance? If this is possible, what is the magnitude ratio at resonance compared to that at very low frequencies? Could this ever be a problem? Explain.

5. Can a driver (controller) ever close the loop around an unstable vehicle (plant) and make the entire closed-loop system stable? Consider this question in light of automobile and motorcycle dynamics.

6. [1]Consider the vehicle shown in Figure 6.1:

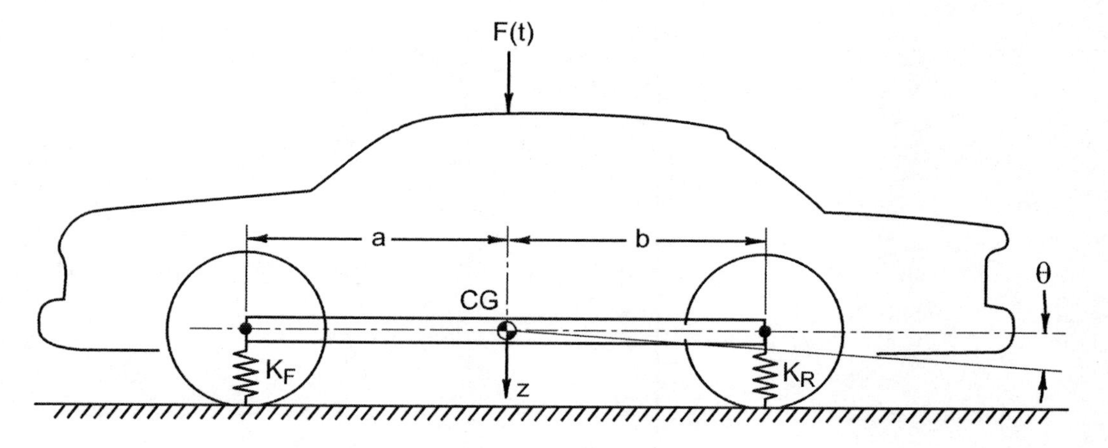

Figure 6.1 *Two degree-of-freedom ride model*

a. Derive the general equations of motion for vertical (heave) and angular (pitch) displacement. For simplicity ignore damping.

b. What is the order of this model?

c. Determine the characteristic equation.

d. Determine the natural frequencies in heave and pitch.

e. Suppose a downward aerodynamic force is exerted directly at the center of gravity. Under what conditions will this produce pure downward translation, without pitch coupling?

f. From the model derived above, determine the front and rear spring deflections as a function of the heave and pitch.

7. The SMD system of RCVD Figure 6.2, p.235, has this characteristic equation:

$$s^2 + 2\zeta\omega_n s + \omega_n^2 = 0$$

[1]This ride model is discussed extensively on pp.288-311 of *Chassis Design: Principles and Analysis* by William F. and Douglas L. Milliken, SAE, R-206, 2002.

Draw a Root-Locus plot with respect to damping ratio. As you vary ζ over the range $0 < \zeta < \infty$ do the factors of the characteristic equation for this simple system move the way you expected them to move?

8. [2]Derive the spring coefficient K_T and the damping coefficient C given in derivative notation by RCVD Table 6.2, p.241:

$$K_T = (aC_F - bC_R) + \frac{\ell^2 C_F C_R}{mV^2}$$

$$C = -N_r - \frac{I_z Y_\beta}{mV}$$

Hint: Start with RCVD Equations 5.10, p.150, and follow the derivation sequence described at the top of RCVD p.250.

9. During turn entry, a vehicle transitions from zero lateral acceleration on the straight to steady-state lateral acceleration at mid-corner. Assume the increase in lateral acceleration is linear with time over this distance. Using a point mass model of the automobile, derive expressions for yaw velocity, path curvature and yaw acceleration as functions of time. Assume vehicle speed remains constant (which approximates, for example, an IRL car in Turn 1 at Indianapolis).

6.2 Simple Measurements and Experiments

NOTE: All experiments should be carried out at low speeds (say, below 30 mph) under safe conditions. A shopping market parking lot on early Sunday morning often makes a good skidpad.

WARNING: The following experiments can produce significant amounts of tire wear in the shoulder area of the tires. Do not continue the experiments for many laps unless you are unconcerned about tire wear. Tire wear can be reduced by performing the experiments on wet or moist surfaces.

You may want to make several preliminary measurements and calculations before beginning these experiments, such as Ackermann steer angle, wheelbase measurement, etc.

[2]This question inspired by Don Girard.

1. Experiment 2 of Chapter 2 proposed a frequency response test. In light of those results and the results from the problems in this chapter, conduct a vehicle frequency response experiment using ten combinations of five input frequencies (0.1, 0.2, 0.5, 1.0 Hz and as fast as you can cycle the steering wheel, probably 2.0 Hz or so), and two input amplitudes (±10 deg. and ±45 deg.). Use a forward velocity of 20 mph for every test. Based on the results of these experiments, is your car linear at this speed over the range of frequencies and amplitudes chosen? Explain.

2. At 15 mph provide a step steering input of 45 deg. at the steering wheel and hold it until your car achieves steady-state and is traveling around a circular path. Repeat the experiment with a steering input of 45 deg. amplitude, but one which linearly ramps up to that amplitude over a 2 sec. period. Qualitatively note the differences in the length of the transient period of operation.

   HINT: It is easier to see such differences if the tests are done on a surface that provides visual tracks, such as a light rain or snow-covered parking lot. Make a sketch of the expected responses before doing the experiment.

   Next, make a device for measuring vehicle path and slip angle by using water fire extinguishers (with reduced nozzle size) on the front and rear bumpers to leave a pair of water paths on the roadway. Repeat the step steer and ramp steer experiments, then survey the road to determine path radius and vehicle slip angle.

3. While traveling at 30 mph in a straight line at steady-state, provide a steering pulse of 15 deg. at the steering wheel. A pulse consists of a "blip" or very short duration input. You may have to practice this several times before you are satisfied that the input approximates a true Dirac delta pulse input. When the transient period has finished, is your vehicle traveling in the same direction as it was before the input? Explain. What is the response of a system when the input is a unit impulse?

   Think in terms of Laplace transformation techniques. The Laplace transform of the Dirac delta function is 1. Estimate the relaxation length of the tires and consider the influence of this on the results of this experiment.

6.3 Problem Answers

1. Draw the free body diagrams with the mass at equilibrium so no gravity (weight) term needs to be included.

   a. For force excitation: $m\ddot{x} + c\dot{x} + kx = F(t)$

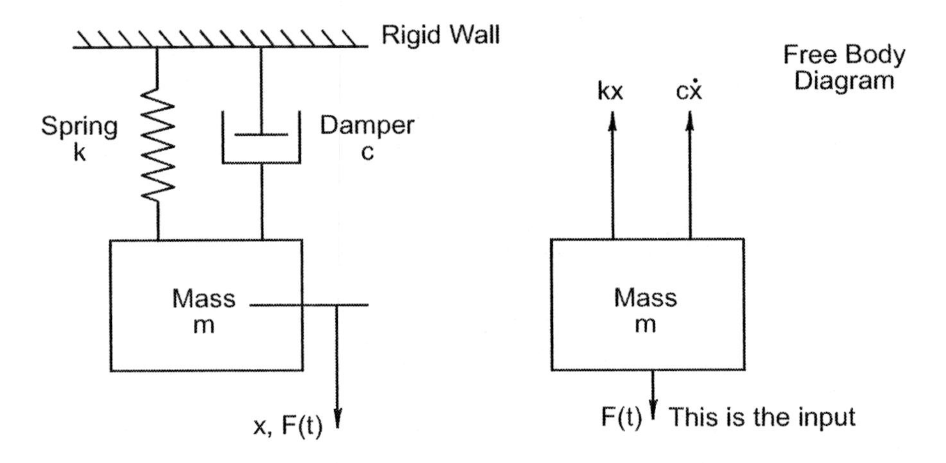

Figure 6.2 *Force excitation sketch and free body diagram*

b. For base excitation:

$$m\ddot{x}+c(\dot{x}-\dot{y})+k(x-y)=0$$

$$m\ddot{x}+c\dot{x}+kx=c\dot{y}+ky$$

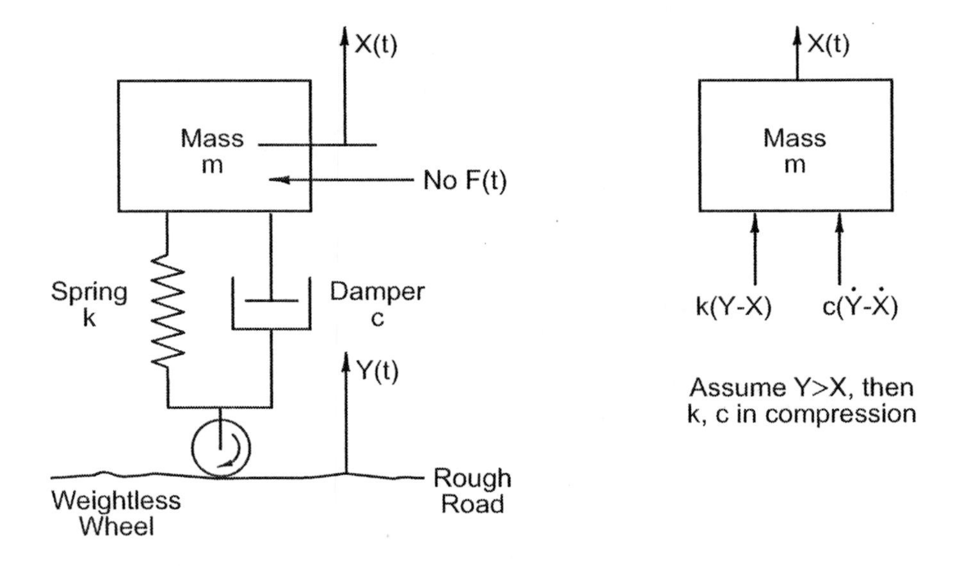

Figure 6.3 *Base excitation sketch and free body diagram*

c. The order is the highest derivative of the dependent variable. This is a second-order model.

d. The characteristic equation (denominator of the transfer function) depends solely on the system itself, not the inputs to the system. As such we would expect either force or base excitation to produce the same characteristic equation. The method of excitation is an input, not a system characteristic.

To calculate the transfer function take the Laplace transform of the equations of motion. By definition:

$$\text{Transfer Function} \equiv \left.\frac{\mathscr{L}(\text{response})}{\mathscr{L}(\text{input})}\right|_{\text{initial conditions}=0}$$

For force excitation:

$$\left(ms^2+cs+k\right)\bar{x}(s)=\bar{F}(s)$$

$$\text{T.F.}=\frac{\bar{F}(s)}{\bar{x}(s)}=\frac{1}{ms^2+cs+k}$$

For base excitation:

$$\left(ms^2+cs+k\right)\bar{x}(s)=(cs+k)\bar{y}(s)$$

$$\text{T.F.}=\frac{\bar{y}(s)}{\bar{x}(s)}=\frac{cs+k}{ms^2+cs+k}$$

Thus, both transfer functions have the same denominator or characteristic equation, namely ms^2+cs+k.

e. The numerator of the transfer function is input-dependent. The two different types of forcing would be expected to have to have different numerators. The derivation of the transfer functions in Problem 1d shows this to be the case. For force excitation the numerator is 1 and for base excitation the numerator is $cs+k$.

2. a. The characteristic equation is of the form:

$$s^2+2\zeta\omega_n s+\omega_n^2=0$$

By direct comparison:

$$\omega_n^2 = 13 \quad \Rightarrow \quad \omega_n = 3.6 \text{ rad./sec.}$$

$$2\zeta\omega_n = 1.4 \quad \Rightarrow \quad \zeta = \frac{1.4}{2\,(3.6)} = 0.19$$

For the damped natural frequency:

$$\omega_d = \omega_n\sqrt{1-\zeta^2} = 3.6\sqrt{1-1.9^2} = 3.53 \text{ rad./sec.}$$

b. The vehicle is stable if the real part of the roots of the characteristic equation are all less than zero. The roots are:

$$s_{1,2} = \frac{1}{2}\left[-1.4 \pm \sqrt{1.4^2 - 4\,(13)}\right] = -0.7 \pm 3.54i$$

The real part is -0.7 which is less than zero. The vehicle is stable.

c. The vehicle is underdamped when $0 < \zeta < 1$. Here $\zeta = 0.19$.

d. Figure 6.4 shows an extended version of RCVD Figure 6.7, p.252. It shows significant overshoot when $\zeta = 0.19$.

e. The plot is shown in Figure 6.5.

3. An asymptotic Bode plot can be drawn from the transfer function. The transfer function can be determined from the equations of motion by taking the Laplace transform. RCVD Equation 5.10 can be rewritten as:

$$m\,(\dot{v} + ru) = (C_F + C_R)\left(\frac{v}{u}\right) + (aC_F - bC_R)\left(\frac{r}{u}\right) - C_F\delta$$

$$I_{zz}\dot{r} = (aC_F - bC_R)\left(\frac{v}{u}\right) + \left(a^2C_F + b^2C_R\right)\left(\frac{r}{u}\right) - aC_F\delta$$

Applying the Laplace transform:

$$m\,(s\bar{v} + \bar{r}u) = (C_F + C_R)\left(\frac{\bar{v}}{u}\right) + (aC_F - bC_R)\left(\frac{\bar{r}}{u}\right) - C_F\delta$$

$$I_{zz}s\bar{r} = (aC_F - bC_R)\left(\frac{\bar{v}}{u}\right) + \left(a^2C_F + b^2C_R\right)\left(\frac{\bar{r}}{u}\right) - aC_F\delta$$

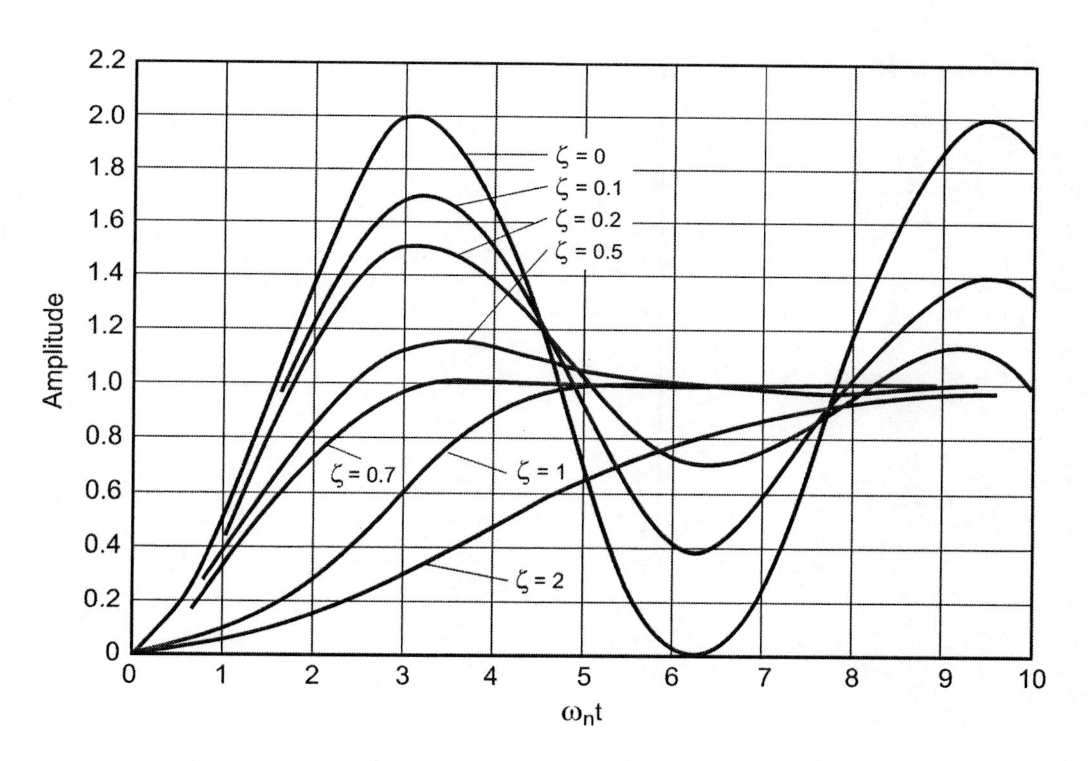

Figure 6.4 *Response for various damping ratios*

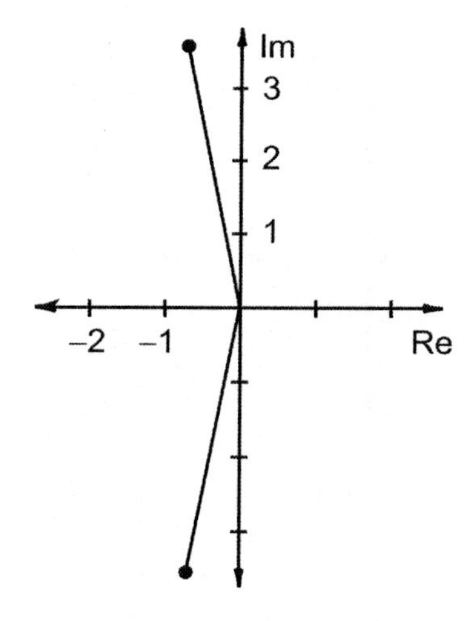

Figure 6.5 *Poles in the s-plane*

These equations are shown in block diagram form in Figure 6.6 (recall that $1/s \sim \int dt$).

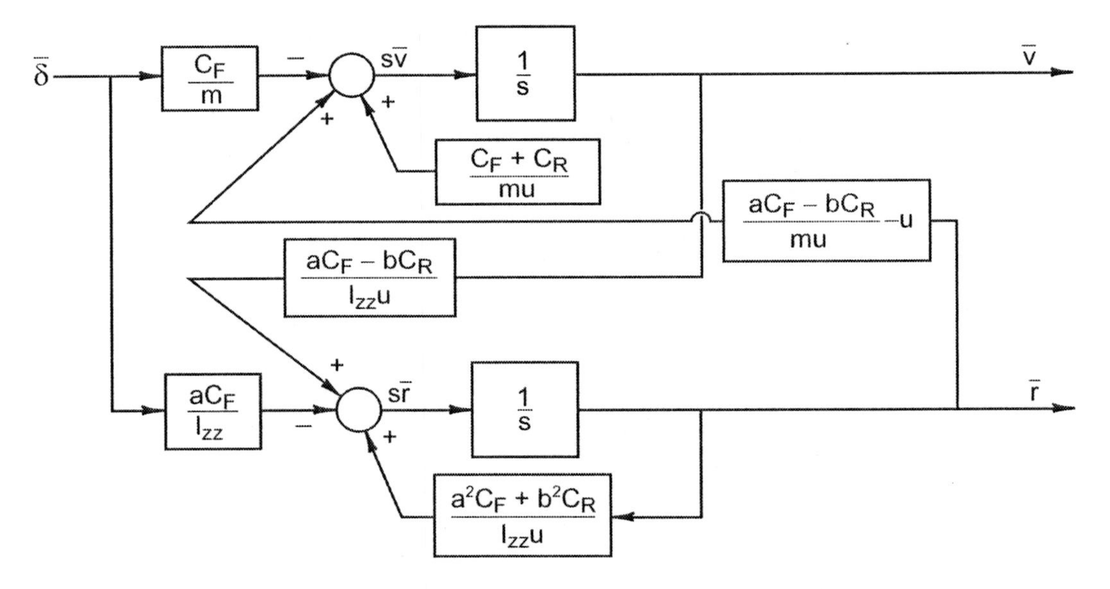

Figure 6.6 *Block diagram of the bicycle model equations of motion*

The pair of equations can be rewritten in matrix form as follows:

$$\begin{bmatrix} ms - \left(\frac{C_F + C_R}{u}\right) & mu - \left(\frac{1}{u}(aC_F - bC_R)\right) \\ -\frac{1}{u}(aC_F - bC_R) & I_{zz}s - \left(\frac{1}{u}\left(a^2C_F + b^2C_R\right)\right) \end{bmatrix} \begin{bmatrix} \bar{v} \\ \bar{r} \end{bmatrix} = \begin{bmatrix} -C_F \\ -aC_F \end{bmatrix} \delta$$

For the vehicle of "First Example" on RCVD p.256 this becomes:

$$\begin{bmatrix} 62.11s + 429.5 & 3271.7 \\ -372.3 & 832s + 7715.5 \end{bmatrix} \begin{bmatrix} \bar{v} \\ \bar{r} \end{bmatrix} = \begin{bmatrix} 13752.0 \\ 33463.2 \end{bmatrix} \delta$$

The determinant of the 2×2 matrix is the characteristic equation of the system (and the denominator of the transfer functions).

$$51675.5s^2 + 836553.7s + 4531861.2 = 0$$

$$s^2 + 16.188s + 87.698 = 0$$

which has roots:

$$s_1, s_2 = -8.094 \pm 4.710i$$

The characteristic equation indicates the natural frequency is 9.36 rad./sec. or 1.49 Hz. The damping ratio is 0.865, which is underdamped.

Returning to the task of determining the yaw rate/steer angle transfer function, it can be found in symbolic form either by hand, matrix algebra or through the use of a computer algebra system. To solve using MATLAB, type:

```
eq1 = 'm*s*v + m*r*u = (CF+CR)*(v/u) + (a*CF-b*CR)*(r/u) - CF*d';
eq2 = 'I*s*r = (a*CF-b*CR)*(v/u) + (a*a*CF+b*b*CR)*(r/u) - a*CF*d';
[v,r]=solve(eq1,eq2)
```

Dividing both v and r by δ results in the transfer functions. The result is rather lengthy and is provided on the accompanying CD. With numerical values, the transfer function is:

$$\frac{\bar{r}}{\bar{\delta}} = \frac{0.84s + 7.57}{s^2 + 16.19s + 87.70} = \frac{0.086(0.111s + 1)}{0.011s^2 + 0.185s + 1}$$

From this the following can be determined:

- DC Gain: $20\log_{10}(0.086) = -21.3$ dB
- Zero at $\omega = 9.09$ rad./sec., slope +20 dB/decade
- Pair of poles at $\omega = 9.36$ rad./sec., slope -40 dB/decade
- Damping ratio: $\zeta = 0.864 < 1$, so there is some (small) amplification at the system natural frequency.

These can be used to draw an asymptotic Bode plot. MATLAB can be used to draw the Bode plot shown in Figure 6.7 with the command:

```
bode([0.84 7.57],[1 16.19 87.7])
```

4. Steering at 1-2 Hz is steering at 6.28-12.56 rad./sec. The natural frequency of the vehicle is 9.36 rad./sec., so it is possible for the driver to operate the vehicle at its natural or resonant frequency.

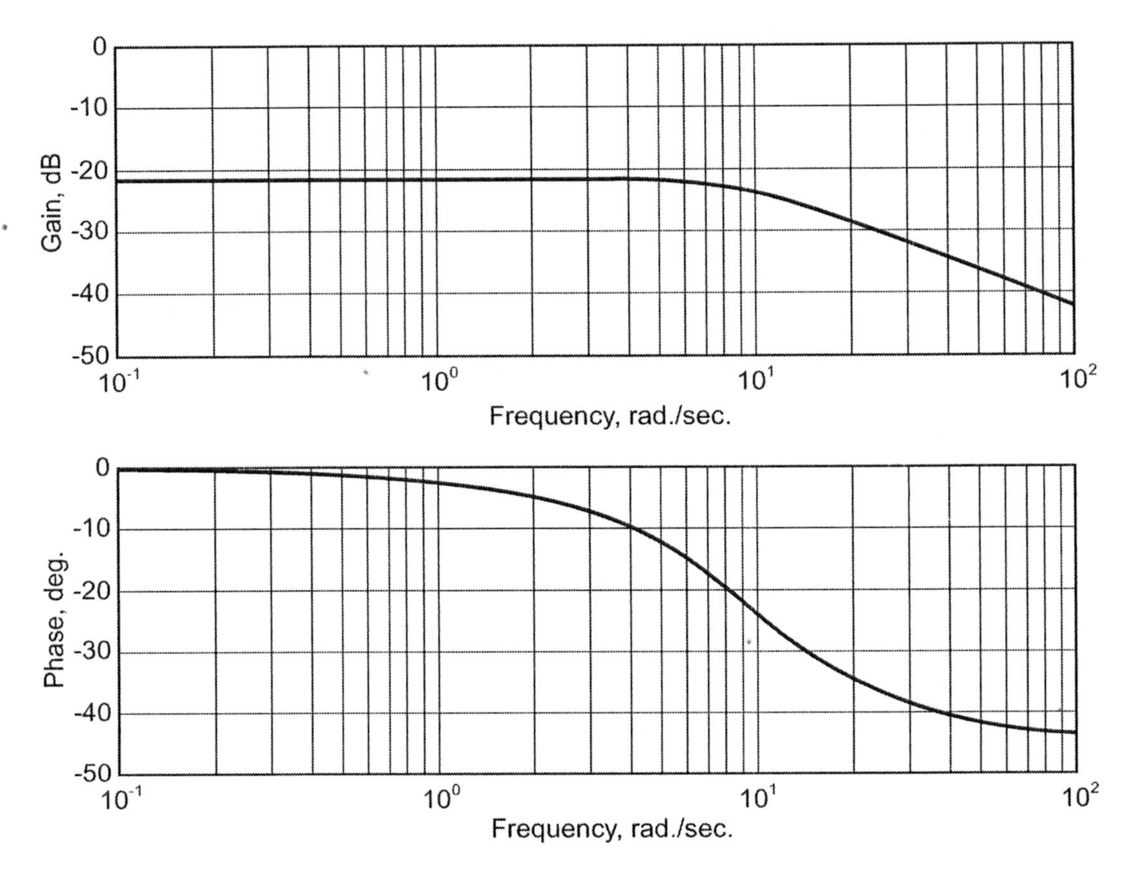

Figure 6.7 *Bode plot for the "First Example" vehicle, RCVD p.256.*

However, as the results of Problem 3 show, the DC gain is approximately −21.3 dB and the gain at resonant frequency is nearly identical. The amplification at resonance is negligible. To the driver, the car would have a normalized r/δ at resonance approximately equal to that at DC, or very low frequency, so nothing unexpected would occur.

Note that while this is true for the simple Bicycle Model, with a more complete model or a real car results may differ (due to the roll mode, tire lag or other factors).

5. It is certainly possible for a driver to operate an unstable vehicle in closed-loop fashion. If it weren't, no bicycle or motorcycle could ever be ridden. Perhaps the most fundamental historical precedents in systems and control theory have been concerned with precisely the problem of stabilizing an unstable plant with a clever controller. See RCVD Figure 5.1, p.124, to see (in block diagram form) just how this is done.

At low speed a bicycle or motorcycle is similar to an upside-down pendulum and is inherently unstable. For a regular pendulum, with its mass located below the pivot, the equation of motion is:

$$\ddot{\phi} + \sqrt{g/\ell}\phi = 0$$

$$s_1, s_2 = \pm i\sqrt{g/\ell}$$

The roots of the characteristic equation have their real part equal to zero, indicating neutral stability. For the inverted pendulum,

$$\ddot{\phi} - \sqrt{g/\ell}\phi = 0$$

$$s_1, s_2 = \pm\sqrt{g/\ell}$$

One root is positive, indicating that the system is unstable in the absence of a controller.

6. a. Summing forces in the heave direction:

$$m\ddot{z} = -K_F(z + a\sin\theta) - K_R(z - b\sin\theta) + F_z(t)$$

Summing pitch moments, where I_y is the pitch inertia:

$$I_y\ddot{\theta} = -K_F a\cos\theta(z + a\sin\theta) + K_R b\cos\theta(z - b\sin\theta) + M_y(t)$$

Linearizing for small pitch angles:

$$m\ddot{z} = -K_F(z + a\theta) - K_R(z - b\theta) + F_z(t)$$

$$I_y\ddot{\theta} = -K_F a(z + a\theta) + K_R b(z - b\theta) + M_y(t)$$

Rearranging to group terms with respect to the motion variables:

$$m\ddot{z} = -(K_F + K_R)z - (K_F a - K_R b)\theta + F_z(t)$$
$$I_y\ddot{\theta} = (-K_F a + K_R b)z - \left(K_F a^2 + K_R b^2\right)\theta + M_y(t)$$

b. This is a fourth-order model, the result of two coupled second-order equations.

c. Start by taking the Laplace transform of the equations of motion:

$$ms^2 Z + (K_F + K_R) Z + (K_F a - K_R b) \Theta = F_z(s)$$

$$I_y s^2 \Theta + (K_F a - K_R b) Z + \left(K_F a^2 + K_R b^2\right) \Theta = M_y(s)$$

In matrix form:

$$\begin{bmatrix} ms^2 + K_F + K_R & K_F a - K_R b \\ K_F a - K_R b & I_y s^2 + K_F a^2 + K_R b^2 \end{bmatrix} \begin{bmatrix} Z \\ \Theta \end{bmatrix} = \begin{bmatrix} F_z(s) \\ M_y(s) \end{bmatrix}$$

The determinant yields the characteristic equation:

$$mI_y s^4 + \left(m\left(K_F a^2 + K_R b^2\right) + I_y (K_F + K_R)\right) s^2 + K_F K_R (a+b)^2$$

d. The roots of the characteristic equation give the natural frequencies. The following MATLAB commands produce the solution:

```
eq1 = 'm*I*x^4 + (m*(KF*a^2+KR*b^2)+I*(KF+KR))*x^2
      + KF*KR*(a+b)^2 = 0'
rts = solve(eq1)
omega1 = sympow(rts(1,:),2)
omega2 = sympow(rts(3,:),2)
```

e. The pitch and heave modes are uncoupled when the off-diagonal terms of the state matrix are zero. Based on the answer to Problem 6c, this means that:

$$K_F a - K_R b = 0 \quad \Rightarrow \quad \frac{K_F}{K_R} = \frac{b}{a}$$

f. Denote the front spring deflection as z_F and the rear spring deflection as z_R. An output equation can be written as follows:

$$\begin{bmatrix} z_F \\ z_R \end{bmatrix} = \begin{bmatrix} 1 & -a \\ 1 & b \end{bmatrix} \begin{bmatrix} z \\ \theta \end{bmatrix}$$

7. Begin by rearranging the characteristic equation:

$$1+\zeta\frac{(2\omega_n)}{(s^2+\omega_n^2)}=0$$

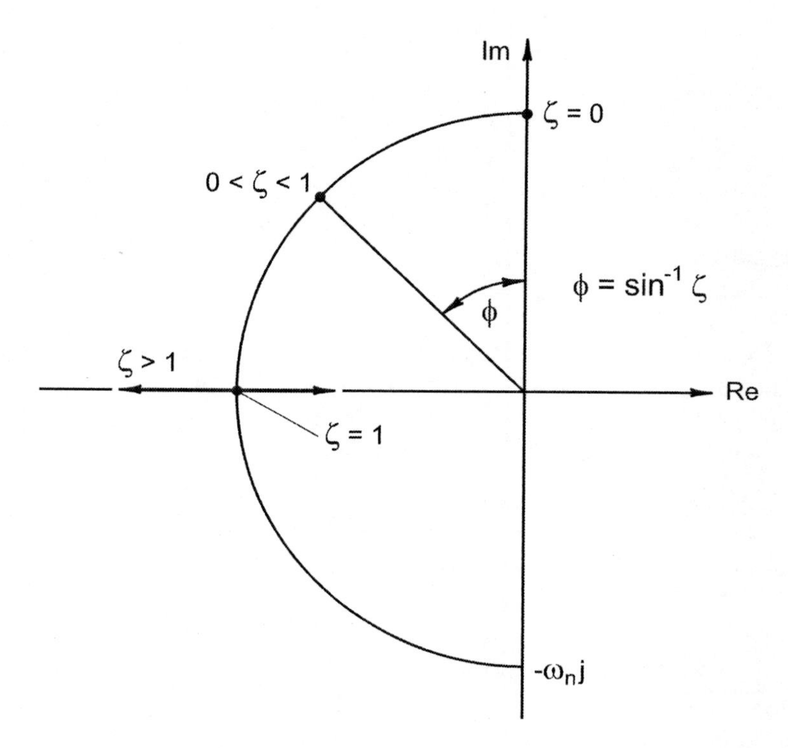

Figure 6.8 *Root-locus plot with respect to damping coefficient*

The following observations can be made:

a. When $\zeta = 0$, the characteristic equation is just $s^2 + \omega_n^2 = 0$, i.e., it has a missing s coefficient. A test of this characteristic equation using a Routh-Hurwitz array would show it to be unstable. Physically, having $\zeta = 0$ means that we have no damping, and that the system would be in perpetual oscillation. The roots of the characteristic equation are at $\pm i\omega_n$, on the imaginary axis of the s-plane. If such a system existed, once started from nonzero initial conditions, it would oscillate sinusoidally forever—an undamped system.

b. When $0 < \zeta < 1$, the characteristic equation has an s coefficient and the roots are at $\sigma \pm i\omega_n$. That is, they have real and imaginary parts. The response is neither on the real axis nor on the imaginary axis. Instead it is in the left hand plane. The system is therefore stable. The natural or free response, if allowed to move from nonzero initial conditions, is for the motion to be a decaying sinusoid—an underdamped system.

c. When $\zeta = 1$, the roots are in the left hand plane, on the real axis and repeated. The response is the sum of two decaying exponential functions with no oscillation, but with identical time constants (i.e., $\tau_1 = \tau_2$). In fact, we could consider a second-order model like this one (i.e., with $\zeta = 1$) to be a pair of repeated first-order models which have identical time constants. We call such a system critically damped ("critical" because $\zeta = 1$ is just enough damping to produce no oscillation).

d. When $\zeta > 1$, the characteristic equation also has two real roots, but now the roots are distinct. Again the response will be the sum of two decaying exponential functions with no oscillation, but now the two time constants of the (considered) first-order systems are different (i.e., $\tau_1 \neq \tau_2$). This system is overdamped because there is more damping than is necessary to eliminate oscillation.

8. The first formula of RCVD Equation 5.10, p.150, is:

$$I_Z\dot{r} = N_\beta\beta + N_r r + N_\delta\delta$$

Solve for β:

$$\beta = \frac{1}{N_\beta}\left(I_Z\dot{r} - N_r r - N_\delta\delta\right)$$

Differentiate with respect to time:

$$\dot{\beta} = \frac{1}{N_\beta}\left(I_Z\ddot{r} - N_r\dot{r} - N_\delta\dot{\delta}\right)$$

Substitute these last two expressions into the second of RCVD Equation 5.10:

$$mV\left(r + \dot{\beta}\right) = Y_\beta\beta + Y_r r + Y_\delta\delta$$

$$mV\left(r + \frac{1}{N_\beta}\left(I_Z\ddot{r} - N_r\dot{r} - N_\delta\dot{\delta}\right)\right) = Y_\beta\frac{1}{N_\beta}\left(I_Z\dot{r} - N_r r - N_\delta\delta\right) + Y_r r + Y_\delta\delta$$

This is a second-order differential equation in yaw rate. It can be rearranged into the familiar form:

$$mVN_\beta r + mV\left(I_z\ddot{r} - N_r\dot{r} - N_\delta\dot{\delta}\right) = Y_\beta\left(I_z\dot{r} - N_r r - N_\delta\delta\right) + Y_r N_\beta r + Y_\delta N_\beta\delta$$

$$I_z\ddot{r} - N_r\dot{r} + N_\beta r - N_\delta\dot{\delta} = \frac{Y_\beta I_z}{mV}\dot{r} + \frac{Y_r N_\beta}{mV}r - \frac{Y_\beta N_r}{mV}r - \frac{Y_\beta N_\delta}{mV}\delta + \frac{Y_\delta N_\beta}{mV}\delta$$

$$I_z\ddot{r} + \left(-N_r - \frac{Y_\beta I_z}{mV}\right)\dot{r} + \left(N_\beta + \frac{Y_\beta N_r - Y_r N_\beta}{mV}\right)r = \left(\frac{Y_\delta N_\beta - Y_\beta N_\delta}{mV}\right)\delta + N_\delta\dot{\delta}$$

From this:

$$\text{Inertia} = I_z$$

$$\text{Damping Coefficient} = -N_r - \frac{Y_\beta I_z}{mV}$$

$$\text{Spring Constant} = N_\beta + \frac{Y_\beta N_r - Y_r N_\beta}{mV}$$

$$\text{Forcing Function} = f(\delta)$$

9. The assumption of a linear increase in lateral acceleration with time says that lateral acceleration can be represented by:

$$A_Y = kt$$

where k is a constant. For a point mass model at constant speed the following relationships apply:

$$A_Y = \frac{V^2}{R} = Vr$$

From this we can develop an expression for yaw rate:

$$r = \frac{A_Y}{V} = \frac{kt}{V} = \left(\frac{k}{V}\right)t$$

Thus, yaw rate varies linearly with time. The proportionality constant is k/V.

To calculate path curvature, ℓ/R, we first develop an expression for R as a function of time:

$$R = \frac{V^2}{A_Y} = \left(\frac{V^2}{k}\right)\frac{1}{t}$$

Substituting into ℓ/R gives:

$$\frac{\ell}{R} = \left(\frac{\ell k}{V^2}\right)t$$

Path curvature varies linearly with time. The proportionality constant is $\ell k/V^2$.

Yaw acceleration is the rate at which yaw rate is changing with respect to time:

$$\dot{r} = \frac{dr}{dt} = \frac{d}{dt}\left(\frac{kt}{V}\right) = \frac{k}{V}$$

When lateral acceleration varies linearly with time, yaw acceleration remains constant. This implies that the yaw moment also remains constant:

$$N = \frac{I_Z k}{V}$$

6.4 Comments on Simple Measurements and Experiments

1. The response will vary depending on the car. Typically the yaw response will peak at some intermediate frequency and then roll off at higher frequencies. There will also be a vehicle roll angle response and, again depending on the particular car, there may be a significant peak or resonance in the roll response.

   It is likely that there will be significant differences between the small steering angle response and the large angle response. With a small steering input on a smooth road, the friction in the suspension and shock absorbers may effectively reduce the vehicle to a go-kart—no suspension motion except in deflections of the tires. This will greatly increase the ride rate (and roll rate) when compared to tests run with a larger steering input. If the suspension doesn't move, the ride rate equals the tire rate.

   If you have access to an empty, wide and smooth road it may be possible to try this experiment at higher speeds. The steering will be more sensitive at higher speeds and the steer angles should be reduced accordingly. At higher speeds a peak (resonance or natural frequency) in the yaw response may be easier to detect and it will be at a different frequency than at low speed because the effect of the tire lag is greatly reduced as speed increases.

2. When a fast step/ramp is applied, you should be able to detect both the tire lag (longer at lower speeds) and the vehicle lag. If the ramp steer is applied slowly the vehicle has time to "catch up" and the lag may be undetectable.

   It is relatively easy to measure the steady-state vehicle slip angle if marks (for example, the water trails described earlier) are left after the car passes. This is probably not precise enough to track the attitude of the vehicle during the transient although it should give qualitative results. Olley describes the "checkerboard" experiment run by Ken Stonex at General Motors in *Chassis Design: Principles and Analysis*[3]. This was perhaps the first attempt (1930s) to measure the path and attitude of a car during a transient. While the instrumentation was primitive by modern standards, the results from these early tests were surprisingly good.

3. While it is not possible to produce a perfect impulse, with some practice a very short input pulse should be possible. In many systems, the Dirac delta function is used because it excites all the system's frequencies (as opposed to, say, a sinusoid which does not). In this case the short duration pulse, even if it is of large amplitude, theoretically has a negligible effect on vehicle response due to the relaxation length of the tires. In practice, the heading of the car might change by a few degrees since the applied pulse is more likely to be a triangle wave input (due to human limitations).

---

[3]See page 224 of *Chassis Design: Principles and Analysis*, Milliken & Milliken, SAE R-206, 2002.

CHAPTER **7**

Steady-State Pair Analysis

Solutions to this chapter's problems begin on page 90.

Unless otherwise stated, the problems in this chapter use the following vehicle data. This vehicle is referred to as the "**CH7**":

- Wheelbase = 104.5 in.
- Total weight = 3700 lb.
- 57/43 Front/Rear weight distribution
- Front track = 61.5 in.
- Rear track = 57.8 in.
- Effective CG height at front track = h_{eF} = 16.5 in.
- Effective CG height at rear track = h_{eR} = 23.5 in.
- Tire characteristics as shown in RCVD Figure 2.10 and repeated in Figure 7.1.

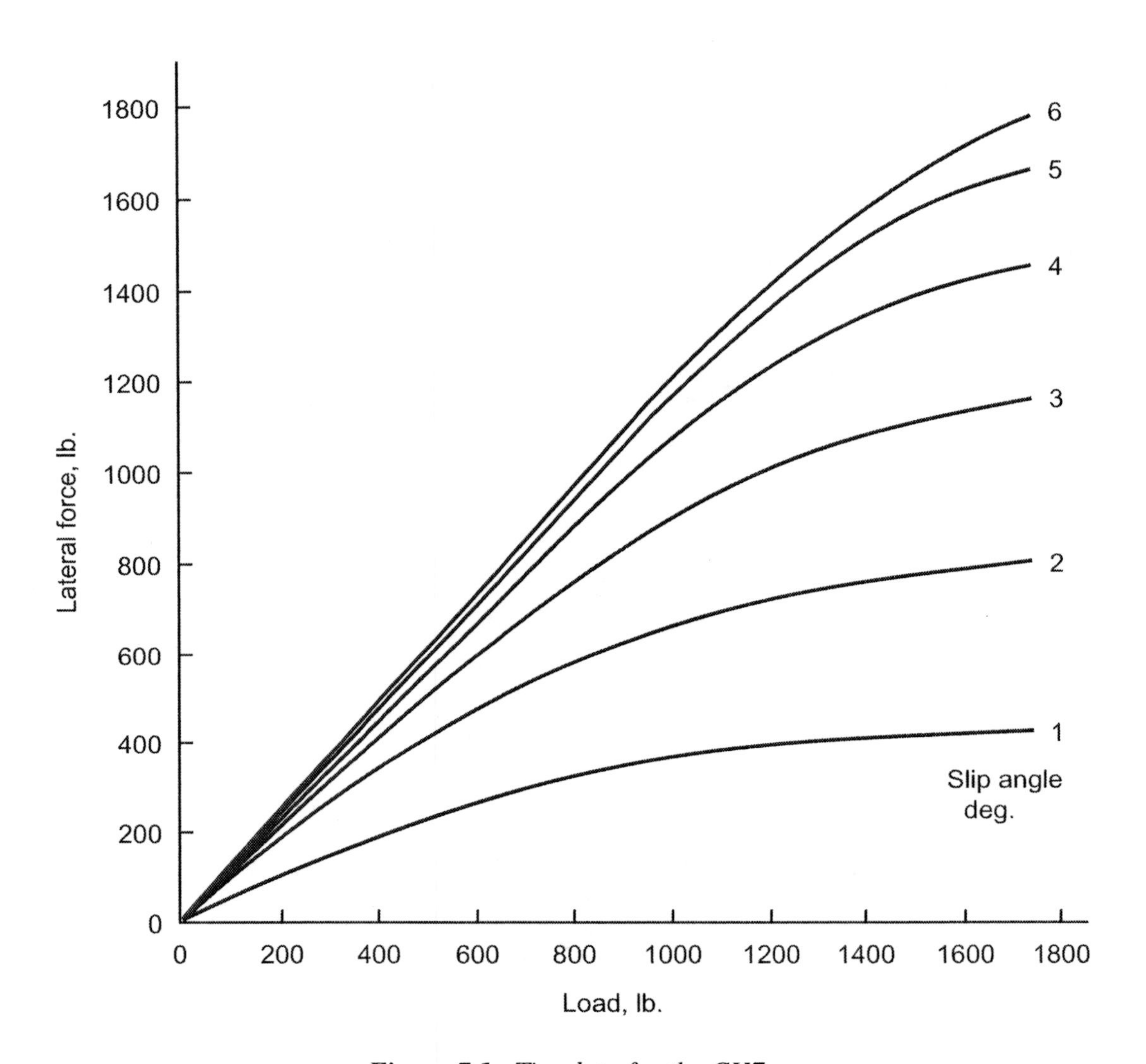

Figure 7.1 *Tire data for the CH7*

7.1 Problems

1. Calculate the static wheel loads for the CH7. Then calculate the dynamic tire normal loads when the vehicle is in a steady-state right hand turn at 0.35g lateral acceleration. At what lateral acceleration does a dynamic normal load reach zero? Which wheel is it?

2. For the CH7, make a plot of available lateral acceleration versus h_e/t for each axle. Assume that the tire/road coefficient of friction is given by the equation $\mu = 1.22 - (0.00024 \times \text{load})$. Examine the figure to determine if the car exhibits terminal plow, spin or drift. Explain your line of reasoning. For this first-order calculation, ignore slip angles.

3. Make plots similar to those shown in RCVD Figure 7.2, p.286, for the CH7 in a right hand turn. Plot:

    a. Lateral acceleration potential for the front track as a function of fraction load transfer

    b. Lateral acceleration potential for the front track as a function of slip angle

    c. Lateral acceleration potential for the rear track as a function of fraction load transfer

    d. Lateral acceleration potential for the rear track as a function of slip angle

    For purposes of these calculations ignore Ackermann geometry, suspension geometry effects, roll axis geometry and slip ratio/friction circle effects.

4. Without making any calculations whatsoever, discuss the effects of the following modifications to a vehicle on its lateral force potential and US/OS/NS characteristics:

    a. Raising or lowering h_{eF} and/or h_{eR}

    b. Widening or narrowing the front and/or rear track

    c. Increasing or decreasing the front and/or rear tire cornering stiffnesses

5. Suspension and steering geometry plays an important role in steady-state pair analysis. Explain how it can affect a vehicle's understeer/oversteer characteristics. Keep your answer brief, as suspension and steering effects will be covered in more detail in Chapters 17 and 19.

6. Suppose an understeering car has a front and rear anti-roll bar. What would you do to move the car toward neutral steer?

    a. Stiffen the front anti-roll bar

    b. Stiffen the rear anti-roll bar

    c. Soften the front anti-roll bar

    d. Soften the rear anti-roll bar

Will more than one of the above strategies achieve the desired handling modification? Explain. Suppose, in addition to neutral steer, it is also desired to make the vehicle have the highest possible lateral cornering limit. Would this additional requirement change your answer above? Consider cases on both smooth and rough roads. Would road roughness influence your decision?

7. Based on your answers to Problem 6, examine the lower left plot of RCVD Figure 7.3, p.290, which addresses the effect of roll stiffness distribution. First, calculate the basic US/OS/NS character of the vehicle described in RCVD Section 7.3, assuming equal tires all around. Then discuss whether this portion of RCVD Figure 7.3 makes sense in light of your results from Problem 6.

8. The formulas presented in this chapter on Pair Analysis do not include roll stiffness distribution. Why? How could the formulas be used or modified to include roll stiffness distribution?

7.2 Problem Answers

1. The CH7 has a 57/43 weight distribution and weighs 3700 lb. The following static wheel loads result, assuming a CG on the vehicle centerline:

$$F_{zLF} = F_{zRF} = \frac{0.57\,(3700)}{2} = 1054.5 \text{ lb.}$$

$$F_{zLR} = F_{zRR} = \frac{0.43\,(3700)}{2} = 795.5 \text{ lb.}$$

To determine the dynamic wheel loads while cornering at 0.35g, start by determining the fraction load transfer across each axle. First the front:

$$FLT = 2A_{yF}\frac{h_{eF}}{t_F} = 2\,(0.35)\,\frac{16.5}{61.5} = 0.188$$

$$FLT = \frac{\Delta F_{zF}}{W_F} \quad \Rightarrow \quad \Delta F_{zF} = (FLT)\,(W_F) = 0.188\,(2109) = 396.5 \text{ lb.}$$

$$F_{zLF} = 1054.5 + \frac{\Delta F_{zF}}{2} = 1252.75 \text{ lb.} \quad \text{and} \quad F_{zRF} = 1054.5 - \frac{\Delta F_{zF}}{2} = 856.25 \text{ lb.}$$

As for the rear:

$$FLT = 2A_{yR}\frac{h_{eR}}{t_R} = 2(0.35)\frac{23.5}{57.8} = 0.285$$

$$FLT = \frac{\Delta F_{zR}}{W_R} \quad \Rightarrow \quad \Delta F_{zR} = (FLT)(W_R) = 0.285(1591) = 452.8 \text{ lb.}$$

$$F_{zLR} = 795.5 + \frac{\Delta F_{zR}}{2} = 1021.9 \text{ lb.} \quad \text{and} \quad F_{zRR} = 795.5 - \frac{\Delta F_{zR}}{2} = 569.1 \text{ lb.}$$

A wheel load will go to zero when the fraction load transfer at one axle equals 1.0. Since the h/t ratio at the rear is higher than at the front, the inside rear wheel will be the first to go to zero load. This will occur at the following lateral acceleration:

$$FLT = 2A_{YR}\frac{h_{eR}}{t_R} \quad \Rightarrow \quad 1.0 = 2(A_{YR})\frac{23.5}{57.8} \quad \Rightarrow \quad A_{YR} = 1.23g$$

The right rear wheel reaches zero normal load at $A_Y = 1.23g$.

2. Define η as the h/t ratio for a given axle. The ratios for the front and rear axles are:

$$\eta_F = \frac{h_{eF}}{t_F} = \frac{16.5}{61.5} = 0.268$$

$$\eta_R = \frac{h_{eR}}{t_R} = \frac{23.5}{57.8} = 0.406$$

Summing the lateral forces gives:

$$WA_Y = F_{zLF}\mu_{LF} + F_{zRF}\mu_{RF} + F_{zLR}\mu_{LR} + F_{zRR}\mu_{RR}$$

Define p as the amount of weight on each axle normalized by the total weight. For the front axle $p = 0.57$ and for the rear axle $p = 0.43$. Then, for either axle:

$$\begin{aligned} pWA_y = &\left[\left(\frac{pW}{2} + \eta pWA_y\right)\left(c_0 - c_1\left(\frac{pW}{2} + \eta pWA_y\right)\right)\right] \\ &+ \left[\left(\frac{pW}{2} - \eta pWA_y\right)\left(c_0 - c_1\left(\frac{pW}{2} - \eta pWA_y\right)\right)\right] \end{aligned}$$

where the first term on the right hand side is the lateral force produced by the outside (heavily loaded) tire and the second term is the lateral force produced by the inside (lightly loaded) tire. The constants c_0 and c_1 correspond to 1.22 and 0.00024, respectively.

This can be simplified (the use of MATLAB, Mathematica, Maple or a calculator with a Computer Algebra System[1] is helpful here):

$$2c_1\eta^2p^2W^2A_y^2 + \frac{c_1p^2W^2}{2} + pWA_y - c_0pW = 0$$

$$\left(2c_1\eta^2pW\right)A_y^2 + A_y + \frac{c_1pW}{2} - c_0 = 0$$

which is quadratic with respect to lateral acceleration. Solving:

$$A_Y = \frac{-1 \pm \sqrt{1 - 4\left(c_1\eta^2pW\right)\left(c_1pW - 2c_0\right)}}{4c_1\eta^2pW}$$

For the front axle substitute $pW = 0.57(3700) = 2109$ lb. and introduce numerical values for c_0 and c_1:

$$A_{YF} = \frac{-1 \pm \sqrt{1 + 3.9153\eta_F^2}}{2.0246\eta_F^2}$$

And for the rear:

$$A_{YR} = \frac{-1 \pm \sqrt{1 + 3.1432\eta_R^2}}{1.5274\eta_R^2}$$

These are plotted in Figure 7.2 for a range of η-values.

Since the rear line is always above the front line, this vehicle plows whenever the front and rear effective CG heights are equal. For the CH7 the following values are read off Figure 7.2.

$$\eta_F = 0.268 \quad \Rightarrow \quad A_{yF} = 0.907g$$

$$\eta_R = 0.406 \quad \Rightarrow \quad A_{yR} = 0.922g$$

While the CH7 still plows, it is much closer to neutral (terminal drift) than it would have been if $\eta_F = \eta_R$.

---

[1] see Appendix B for a description of useful programs

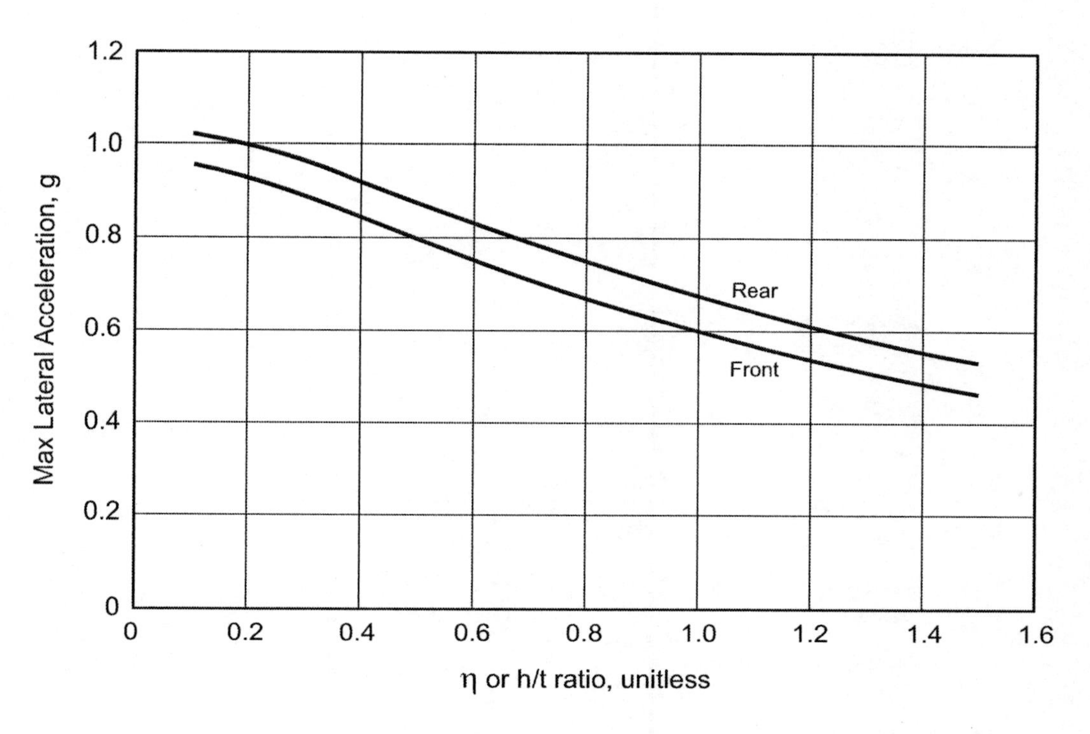

Figure 7.2 *Sustainable lateral acceleration vs. h/t ratio*

3. a. To draw this diagram two sets of lines are needed. The first (horizontal) are at constant slip angle. The second (radial) are at constant h_e/t ratio.

The weight on the front axle is $W_F = 0.57(3700) = 2109$ lb. Each front wheel has a static load of 1054.5 lb. Fraction load transfer is calculated from:

$$FLT = \frac{F_{zLF} - F_{zRF}}{W_F} = \frac{\Delta F_{zF}}{W_F}$$

For a right hand turn:

FLT –	ΔF_{zF} lb.	$\Delta F_{zF}/2$ lb.	F_{zLF} lb.	F_{zRF} lb.
0.0	0.0	0.0	1055	1055
0.2	421.8	211	1266	844
0.4	843.6	422	1477	633
0.6	1265.4	633	1688	422
0.8	1687.2	844	1899	211
1.0	2109.0	1055	2109	0

The above wheel loads are then used with Figure 7.1 to determine the lateral force produced by each wheel at a given slip angle. These forces are summed and the lateral acceleration for the track is determined from the equation:

$$A_{YF} = \frac{(F_{yLF} + F_{yRF}) \cos\alpha}{W_F}$$

For example, at $\alpha = 2$ deg.:

FLT –	F_{yLF} lb.	F_{yRF} lb.	$F_{yLF} + F_{yRF}$ lb.	A_{YF} g
0.0	650	650	1300	0.616
0.2	700	573	1273	0.604
0.4	735	470	1205	0.571
0.6	760	332	1092	0.518
0.8	770	175	945	0.448
1.0	777	0	777	0.368

This can be repeated at $\alpha = 4$ deg. and $\alpha = 6$ deg. The radial lines employ the following equation:

$$FLT = 2A_{YF}\frac{h_{eF}}{t_F}$$

The following FLT table can be generated by selecting values for A_{YF} and h_{eF}/t_F:

h_{eF}/t_F	A_{YF}				
–	0.2g	0.4g	0.6g	0.8g	1.0g
0.2	0.08	0.16	0.24	0.32	0.4
0.4	0.16	0.32	0.48	0.64	0.8
0.6	0.24	0.48	0.72	0.96	1.2

A value greater than 1.0 for FLT indicates an "inverse load" on the tire (i.e., the road pulling down on the tire instead of pushing up on it). This is not physically possible, so the plot is truncated at 1.0g.

The resulting diagram is shown in Figure 7.3. The individual left and right wheel contributions for the $\alpha = 2$ deg. case are shown as dashed lines.

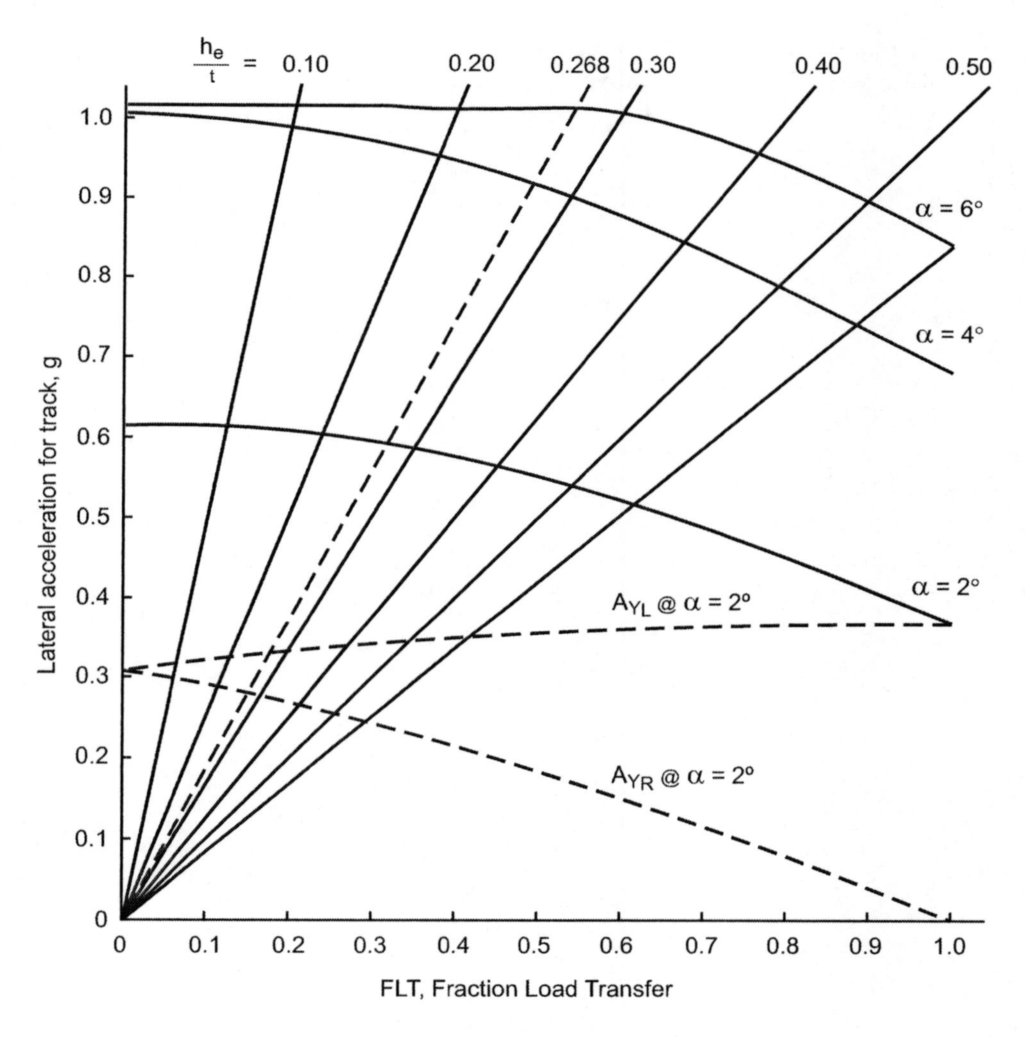

Figure 7.3 *Front potential diagram*

For the pair of front tires on the CH7:

$$\frac{h_{eF}}{t_F} = \frac{16.5}{61.5} = 0.268$$

$$FLT = 2A_{YF}\frac{h_{eF}}{t_F} = 0.536A_{YF}$$

This line is also plotted on Figure 7.3.

b. To plot the lateral acceleration versus slip angle, take the intersection points of the $h_{eF}/t_F = 0.268$ line with each slip angle in Figure 7.3.

α_F deg.	A_{YF} g
2.0	0.590
4.0	0.916
6.0	1.120

Compared with no lateral weight transfer, $h_{eF}/t_F = 0.0$:

α_F deg.	A_{YF} g
2.0	0.610
4.0	1.000
6.0	1.160

These are plotted in Figure 7.4.

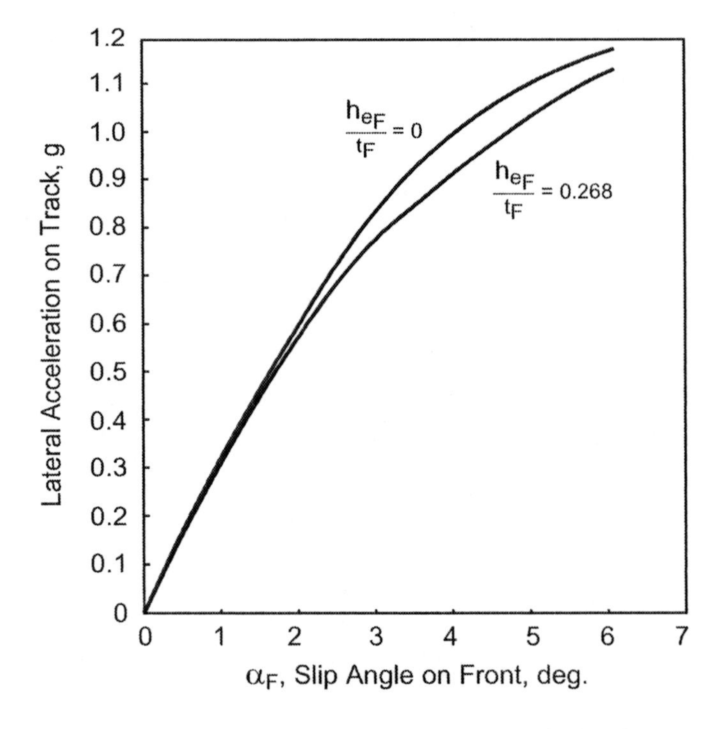

Figure 7.4 *Lateral acceleration vs. slip angle for front track*

c. Calculations are similar to those in Question 3a, only now the rear static weight, CG height and track width are used.

d. Calculations are similar to those in Question 3b with values for the rear of the vehicle.

4. All three of the modifications given will change the amount of lateral force available on one axle relative to the other. Of course, this will change the vehicle US/OS/NS characteristics.

   a. Raising or lowering the effective CG height at the axle (h_{eF} and/or h_{eR}) will change the moment arm through which the centrifugal force acts. If the CG is raised the moment arm will be increased and thus there will be more fraction load transfer on that axle for a given lateral acceleration. Typically, when weight transfer is increased, the total lateral force capability of that axle is diminished (except for high negative camber!).

   b. Expanding on the answer to Part 4a, note that it is really the h/t ratio that matters in calculating fraction load transfer. If this ratio is decreased (either by increasing the track or decreasing the CG height) then there will be less fraction load transfer at a given lateral acceleration, resulting in a smaller reduction in available lateral force from that axle pair.

   c. Increasing or decreasing the cornering stiffness (through changes in tire construction or tire pressure) will increase or decrease, respectively, the available lateral force from a given axle at a given slip angle. So, if the cornering stiffnesses of the front tires are reduced, lateral force is subtracted from the front axle relative to the rear axle. Keeping in mind that the front tires turn the vehicle while the rear tires resist vehicle turning, it follows that decreasing the front tire cornering stiffnesses increases understeer, while decreasing the rear tire cornering stiffnesses increases oversteer. Conversely, increasing the front tire cornering stiffnesses increases oversteer while increasing the rear tire cornering stiffnesses increases understeer.

5. The suspension and steering geometry controls how the tires are presented to the road. The layout establishes the steer and camber angles of each wheel at static ride position and controls how these angles change as a function of ride position and steering wheel angle. It also determines the relative changes in these angles between two wheels on a given axle (e.g., Ackermann steering or roll steer). Compliances will further modify the steer and camber angle of each wheel. The net result is that, by controlling the tire's operating condition (α and γ), the suspension and steering controls the lateral force produced by a pair of wheels, directly influencing US/OS/NS characteristics.

6. An anti-roll bar controls the amount of lateral weight transfer that occurs when the wheels on a given axle are displaced vertically by differing amounts (whether by hitting a bump with one wheel or by cornering with subsequent body roll). Recall that tire load sensitivity is such that weight transfer reduces the lateral force capability of an axle.

   Stiffening an anti-roll bar will reduce the lateral force capability of an axle. If a car understeers, it produces too much lateral force at the rear axle when compared to the front axle. Therefore the rear anti-roll bar could be stiffened (which would reduce the lateral force capability of the rear axle) or the front anti-roll bar could be softened (which would increase the lateral force capability of the front axle).

   Increasing the front and rear anti-roll bars such that the roll stiffness *distribution* remains unchanged will not have any effect on the total lateral force capability of the vehicle on a smooth road. Lateral load transfer will remain unchanged, as it is determined by the ratio of CG height to track width. On a rough road, however, large anti-roll bars can be detrimental in cases of one wheel bump. For rough roads, low roll stiffness is desirable to keep the tires in contact with the roadway and maintain lateral force capability.

7. For this vehicle, $aC_F - bC_R = C(a-b)$, ignoring tire load sensitivity effects. Because the vehicle has equal tires all around and $a < b$, it understeers ($N_\beta < 0$). The lower left plot of RCVD Figure 7.3 makes sense. If the roll couple distribution is increased from the nominal value of 62% front to 82% front, the lateral force capability of the front axle is reduced, making the car understeer even more. If the front roll couple is decreased to 42% front, the available lateral force on the front axle is increased and the car becomes less understeer. These trends are consistent with the answers to Problem 6.

8. The load transferred from the two inside wheels to the two outside wheels during steady-state cornering depends on the lateral acceleration, the CG height and the track width. When considering the whole car, roll stiffness has no significant effect on the magnitude of load transferred from the inside to the outside wheels. Roll stiffness distribution does, however, change the relative amounts of load transferred across the front and rear axle. This is covered in RCVD Chapter 18.

   Pair analysis takes the above "whole car" approach to weight transfer and applies it to only one end of the car. In doing so it applies the same equations, only now it makes use of an "effective CG height at the track". Pair analysis is intended to be a simple procedure, amenable to hand calculations, which gives reasonable results. Roll stiffness distribution *could* be included in the pair analysis procedure by raising or lowering the effective CG accordingly, but at this level the techniques of RCVD Chapter 18 are more applicable.

CHAPTER **8**

Force-Moment Analysis

Solutions to this chapter's problems begin on page 103.

8.1 Problems

1. Label the following on Figure 8.1:

    a. Front and rear construction lines

    b. Front tire limit, rear tire limit

    c. Left and right hand turn

    d. Trim line

    e. Maximum trimmed lateral acceleration

    f. Maximum lateral acceleration

    g. Terminal drift/spin

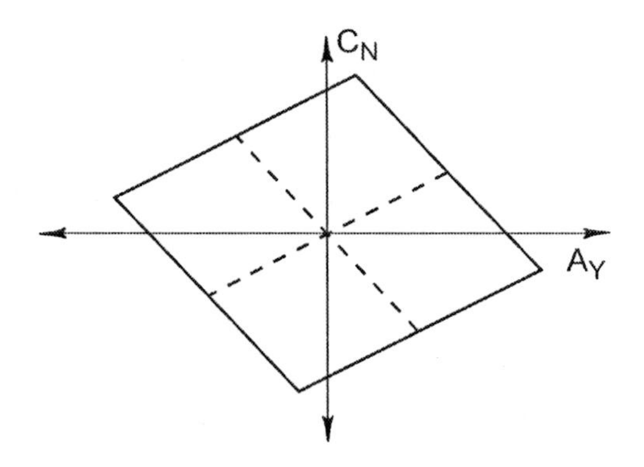

Figure 8.1 *Simplified force-moment diagram*

2. Consider the bicycle model of the automobile with a mass of 60 slugs traveling at a velocity of 30 mph. The vehicle's tires are modeled according to Figure 8.2 where $\alpha_{max} = \pm 6$ deg. Use this data and the vehicle configurations below to construct C_N-A_Y diagrams. Describe each as terminal plow, spin or drift.

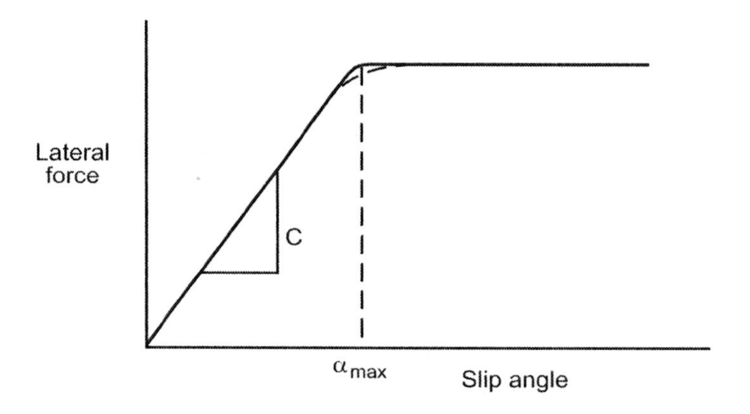

Figure 8.2 *Linear tire with breakaway slip angle*

a. a = 5 ft., b = 5 ft., $C_F = C_R = -420$ lb./deg.

b. a = 6 ft., b = 4 ft., $C_F = C_R = -420$ lb./deg.

c. a = 4 ft., b = 6 ft., $C_F = C_R = -420$ lb./deg.

d. a = 5 ft., b = 5 ft., $C_F = -504$ lb./deg., $C_R = -336$ lb./deg.

e. a = 5 ft., b = 5 ft., $C_F = -336$ lb./deg., $C_R = -504$ lb./deg.

3. Consider a typical MRA Moment Method diagram at a constant speed of 60 mph as shown in Figure 8.3 (this is RCVD Figure 8.7, p.307).

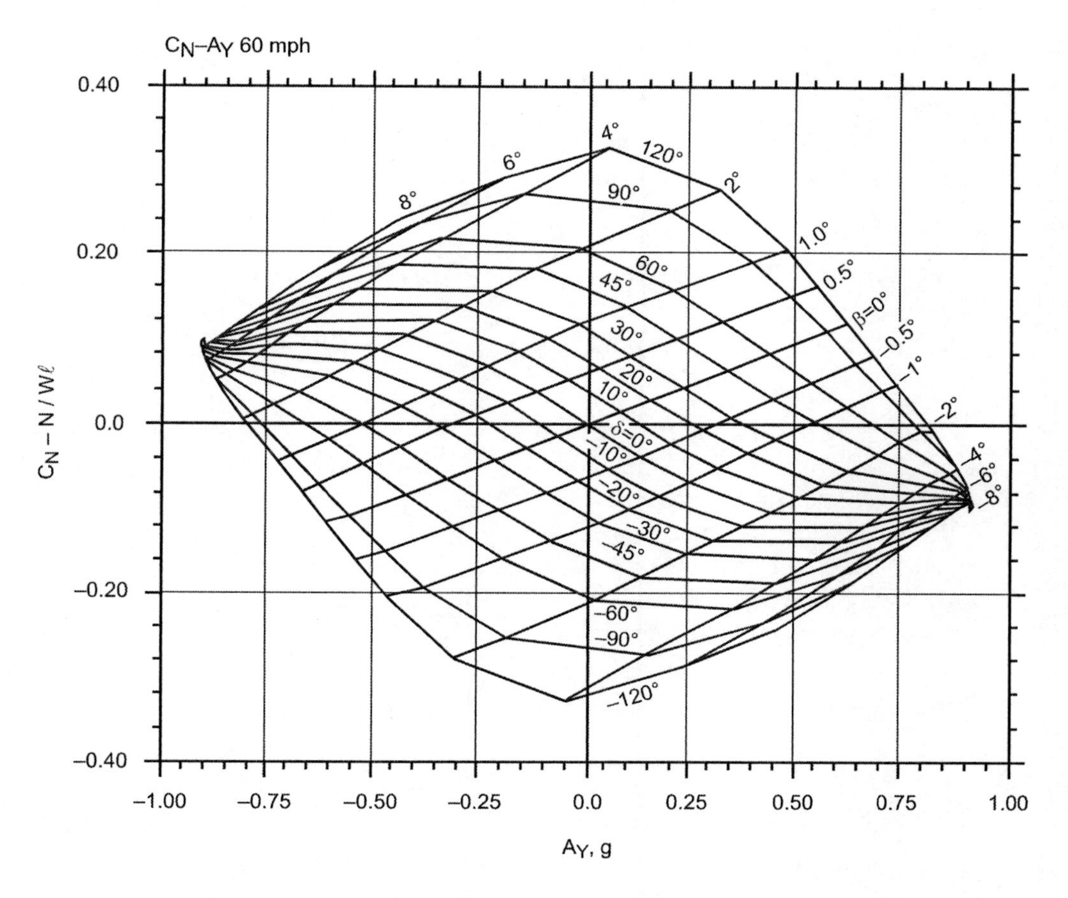

Figure 8.3 *Typical four quadrant MMM diagram*

Perform the following calculations:

a. Determine the static directional stability at $A_Y = 0.3g$ and $A_Y = 0.7g$.

b. Determine the maximum sustainable lateral acceleration, the nature of the limit behavior and a numeric measure of the limit behavior.

c. Determine the steering sensitivity measured at the steering wheel at lateral accelerations $A_Y = 0.3g$ and $A_Y = 0.7g$.

d. Determine the sideslip sensitivity in deg./g at $A_Y = 0.3g$ and $A_Y = 0.7g$.

e. Determine the understeer gradient in deg./g of front wheel steer angle at $A_Y = 0.3g$ and $A_Y = 0.7g$. Use a steering ratio of 20:1 to convert from steering wheel angle to front wheel steer angle. Additionally, this vehicle has an 8.5 ft. wheelbase. (Hint: See RCVD Equation 5.69, p.214.)

f. Plot a curve of steering wheel angle versus lateral acceleration for this car and plot the Ackermann angle (neutral steer response) on the same graph. Measure the slopes at $A_Y = 0.3g$ and $A_Y = 0.7g$, convert to understeer gradient and compare to the results in 3e.

g. Discuss the probable contributors to the understeer gradient at $A_Y = 0.3g$ and $A_Y = 0.7g$ for this car. Does it make sense to calculate understeer at high lateral accelerations in the nonlinear part of the tire performance?

4. It is instructive to trace the path of a transient maneuver on the force-moment diagram. Consider the vehicle of Figure 8.3 initially traveling straight ahead on a flat roadway at a constant speed of 60 mph. Suppose a 180 deg./sec. ramp steer is applied at the steering wheel for 0.25 sec. and held until steady-state cornering is reached. Then, the steer input is removed at the same rate and held at zero until steady-state (straight ahead) is once again achieved. Trace the path of the vehicle state on the force-moment diagram of Figure 8.3.

5. a. Consider a car described by the following measurements:

   - Weight, W = 1800 lb.
   - Wheelbase, $\ell = 8$ ft.
   - Radius of Gyration, k = 3 ft.
   - Yaw Inertia, $I_Z = 503.1$ slug-ft.^{2}

   This car is entering Turn One at Indianapolis:

   - Turn radius at apex = 840 ft.
   - Distance from end of straight to apex = 600 ft.
   - Vehicle speed, V = 204.5 mph and is constant (due to high downforce)
   - Bank angle = 0 (this is a simplification from Indy where the banking at the apex is 9.2 deg.

   Assume the lateral acceleration builds linearly with time from the end of the straight to the apex where it reaches a peak of 3.33g. Calculate the yaw moment, N, during this interval (Hint: See Problem 9 of Chapter 6). Also determine the yaw moment coefficient C_N.

b. Repeat the scenario of 5a, but increase the radius of gyration to 4 ft. Recalculate N and C_N. Compare the result with the result of 5a. Would you expect that the difference in response between the two cars would be very noticeable to a driver?

c. Typically a step steer of 1 deg. at the front wheels for this type of vehicle would result in a yawing moment coefficient of about 0.10 (see RCVD p.312). Compare this value with the C_N values calculated in 5a and 5b above. What does this say about the driver's approach to Turn One at Indianapolis? Taking into account the packaging problems in an IRL car, would you feel it worthwhile to give primary design consideration to reducing yaw inertia?

8.2 Problem Answers

1. The points A-L have been added on Figure 8.4.

a. The front construction line goes from E to G. The rear construction line goes from B to K.

b. The front tire limit is the line from C to L as well as the line from A to J. The rear tire limit goes from A to C as well as from J to L.

c. A left hand turn occurs anywhere on the diagram to the left of the $A_Y = 0$ axis. Right hand turns occur on the diagram to the right of the axis.

d. The trim line extends along the $C_N = 0$ line from D to H.

e. Maximum trimmed lateral acceleration occurs at points D and H.

f. Maximum lateral acceleration occurs at the points A and L.

g. Terminal drift/spin is indicated by the location of points A and L. If L is above the trim line the vehicle has terminal spin, on the trim line it is terminal drift and below this trim line (as in this case) it is terminal plow. If A is above the trim line (as in this case) the vehicle is terminal plow, on the trim line it is terminal drift and below the trim line it is terminal spin.

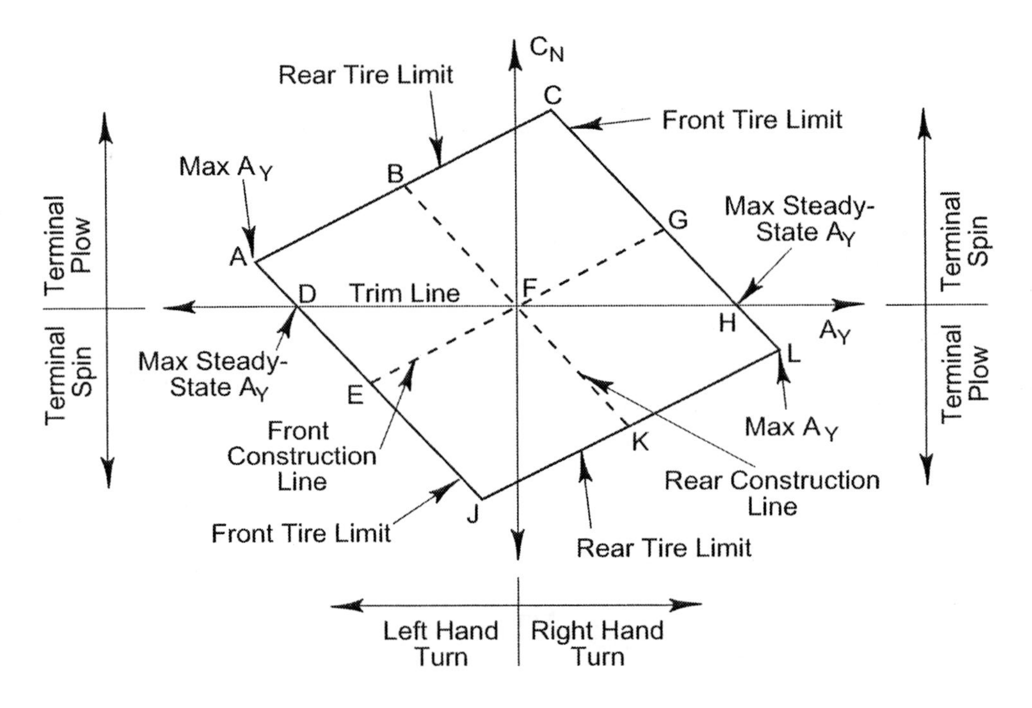

Figure 8.4 *Simplified force-moment diagram*

2. The solutions involve the following calculations.

- Vehicle weight, $W = mg = (60)(32.2) = 1932$ lb.
- Weight times wheelbase, $W\ell = (1932)(10) = 19320$ lb.-ft.
- Maximum lateral acceleration produced by the front tires:

$$A_Y = \frac{(F_{yF})_{max}}{W} = \frac{C_F \alpha_{max}}{W}$$

- Maximum yaw moment coefficient produced by the front tires:

$$C_N = \frac{a \times (F_{yF})_{max}}{W\ell}$$

- Maximum lateral acceleration produced by the rear tires:

$$A_Y = \frac{(F_{yR})_{max}}{W} = \frac{C_R \alpha_{max}}{W}$$

- Maximum yaw moment coefficient produced by the rear tires:

$$C_N = \frac{-b \times (F_{yR})_{max}}{W\ell}$$

These equations determine the endpoints of the front and rear construction lines. From these construction lines the force-moment diagram boundary can be drawn.

a. For the front construction line endpoints, $A_Y = \pm 1.30g$ and $C_N = \pm 0.65$. For the rear construction line endpoints, $A_Y = \pm 1.30g$ and $C_N = \mp 0.65$. Use these points to produce Figure 8.5a which portrays terminal drift.

b. For the front construction line endpoints, $A_Y = \pm 1.30g$ and $C_N = \pm 0.78$. For the rear construction line endpoints, $A_Y = \pm 1.30g$ and $C_N = \mp 0.52$. Use these points to produce Figure 8.5b which portrays terminal plow.

c. For the front construction line endpoints, $A_Y = \pm 1.30g$ and $C_N = \pm 0.52$. For the rear construction line endpoints, $A_Y = \pm 1.30g$ and $C_N = \mp 0.78$. Use these points to produce Figure 8.5c which portrays terminal spin.

d. For the front construction line endpoints, $A_Y = \pm 1.56g$ and $C_N = \pm 0.78$. For the rear construction line endpoints, $A_Y = \pm 1.04g$ and $C_N = \mp 0.52$. Use these points to produce Figure 8.5d which portrays terminal spin.

e. For the front construction line endpoints, $A_Y = \pm 1.04g$ and $C_N = \pm 0.52$. For the rear construction line endpoints, $A_Y = \pm 1.56g$ and $C_N = \mp 0.78$. Use these points to produce Figure 8.5e which portrays terminal plow.

3. a. The static directional stability (or Stability Index, SI) is the amount of restoring moment, N (or coefficient, C_N), per unit disturbance in lateral acceleration, A_Y, at constant steer angle (i.e., fixed control). This slope is taken along the trim line, $C_N = 0$. The slope is negative for stability, meaning that the vehicle has a yawing moment trying to return it to its original trim. The slopes at $A_Y = 0.3g$ and $A_Y = 0.7g$ are shown on Figure 8.6. They are -0.25 and -0.40 C_N/g, respectively.

   b. The maximum sustainable (trimmed) lateral acceleration is measured on the trim line ($C_N = 0$) at the diagram limit. This is shown on Figure 8.7. The answer is 0.82g. This occurs at $\delta = 120$ deg. and $\beta = -2.0$ deg.

   The vehicle has limit understeer (terminal plow), since the front wheels reach their limit first and the rear tires are still producing additional yawing moment with increasing vehicle sideslip β.

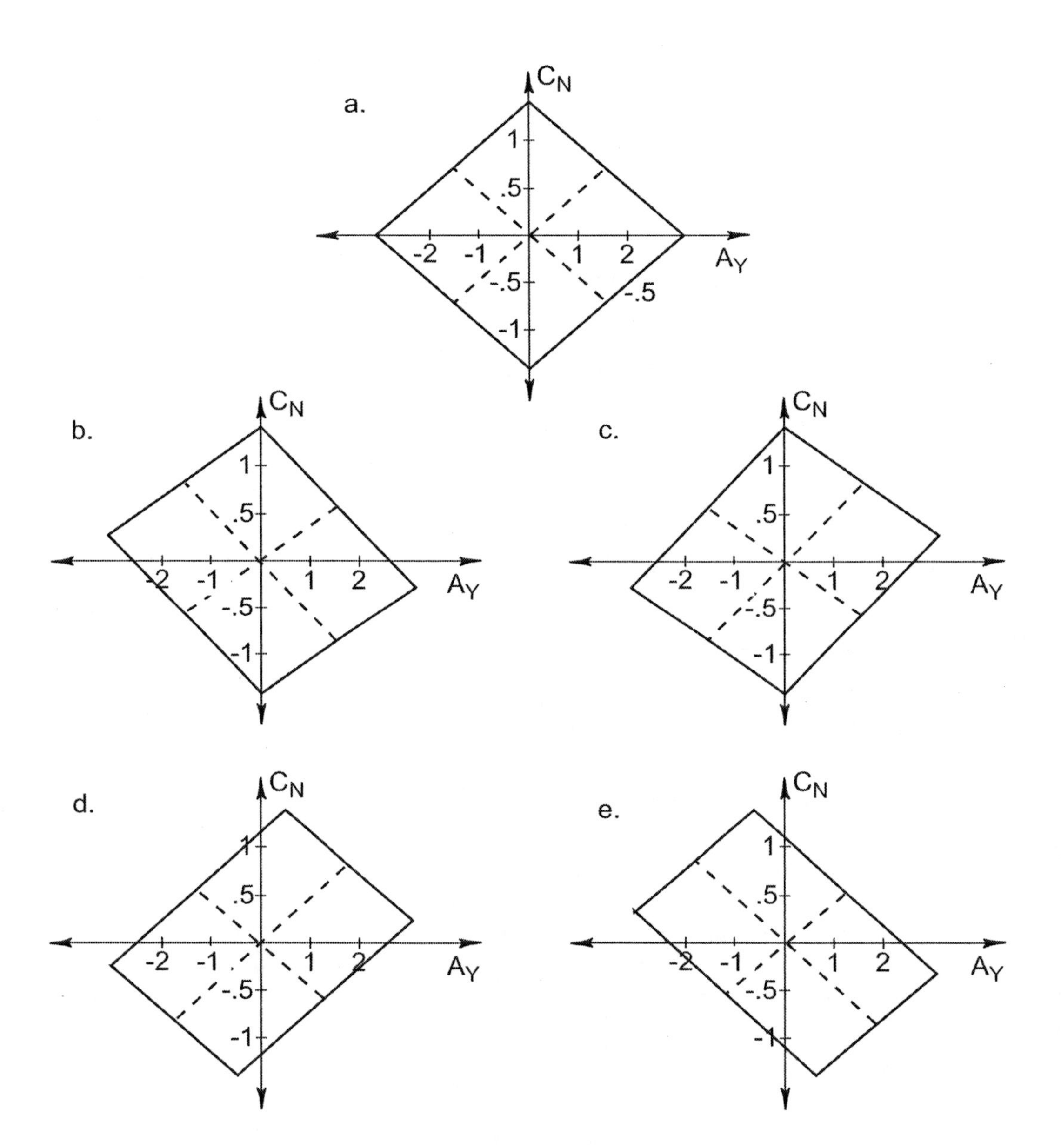

Figure 8.5 *Force-moment diagrams for Problem 2.*

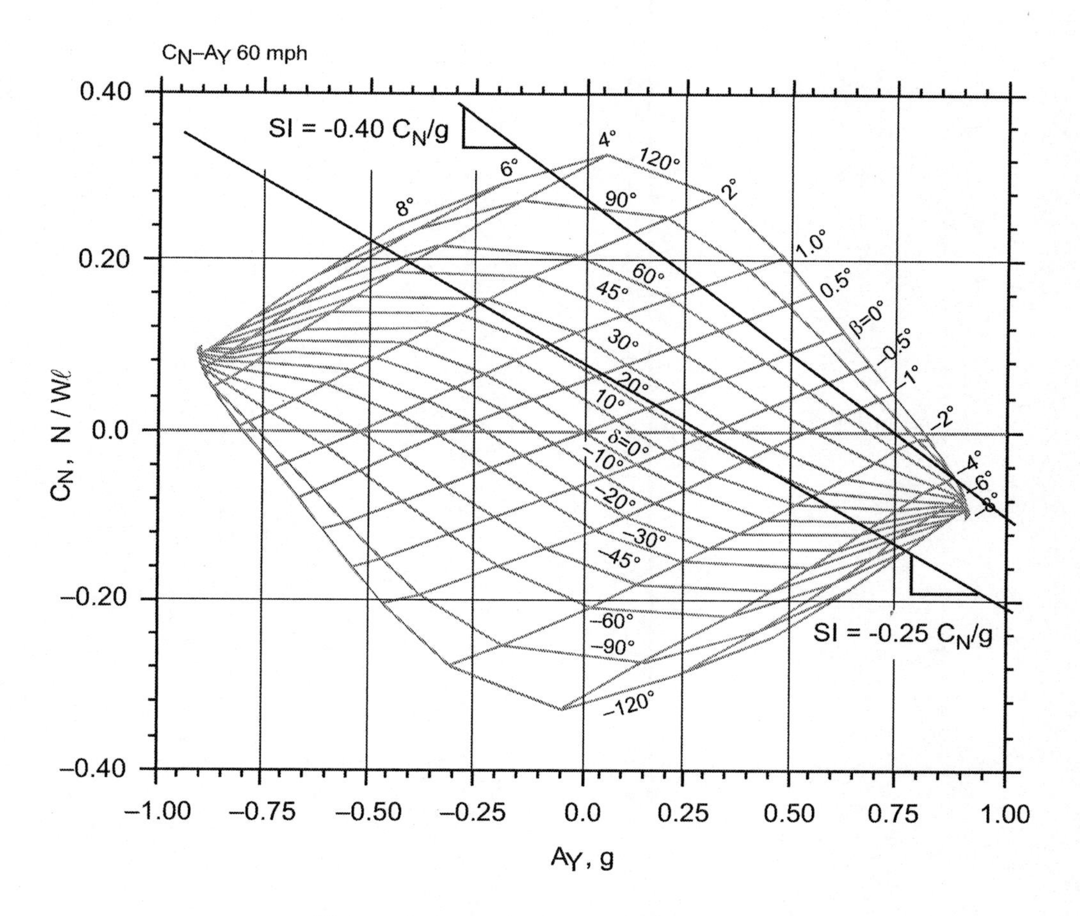

Figure 8.6 *Static directional stability*

Two numeric measures of the limit behavior are:

- Distance a = −0.09, the distance from the horizontal axis (trim line) to the intersection of the front and rear limits.
- Distance b = −0.125, the maximum residual yawing moment from the rear when the front tires are saturated. This is related to the measure of driver confidence used by Peter Wright[1].

c. SAE J670e, *Vehicle Dynamics Terminology*, defines Steering Sensitivity as "the change in steady-state lateral acceleration on a level road with respect to change in steering wheel angle at a given trim and test condition". Industry practice defines it as the change in steady-state lateral acceleration per 100 degrees of steering wheel motion, i.e., g/100 deg. This is measured on a C_N-A_Y diagram

[1]*Moment Method—A Comprehensive Tool for Race Car Development*, SAE paper 942538, abstract on CD.

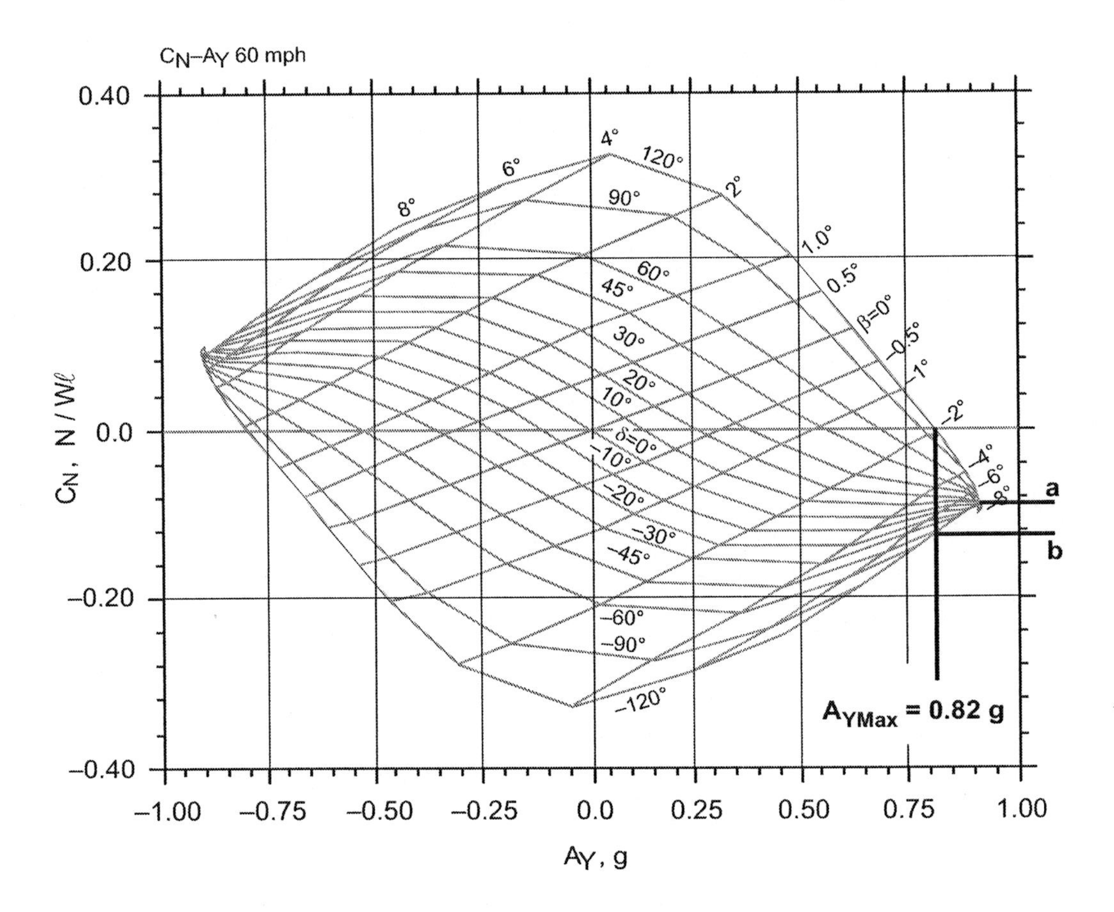

Figure 8.7 Trimmed lateral acceleration and limit behavior

by noting the change in A_Y (i.e., ΔA_Y) corresponding to the intersection of the two steer angle lines (on either side of trim) with the horizontal axis, see Figure 8.8. The change in δ_{SW} is $\Delta\delta_{SW}$ and the steering sensitivity is:

$$\text{Steering Sensitivity} = 100 \times \frac{\Delta A_Y}{\Delta \delta_{SW}}$$

At $A_Y = 0.3\text{g}$, Steering Sensitivity $= 100 \times (0.13/10) = 1.30$ g/100 deg. SW.

At $A_Y = 0.7\text{g}$, Steering Sensitivity $= 100 \times (0.10/30) = 0.33$ g/100 deg. SW.

d. The sideslip sensitivity is expressed as:

$$\text{Sideslip Sensitivity} = \left.\frac{\Delta\beta}{\Delta A_Y}\right|_{A_Y}$$

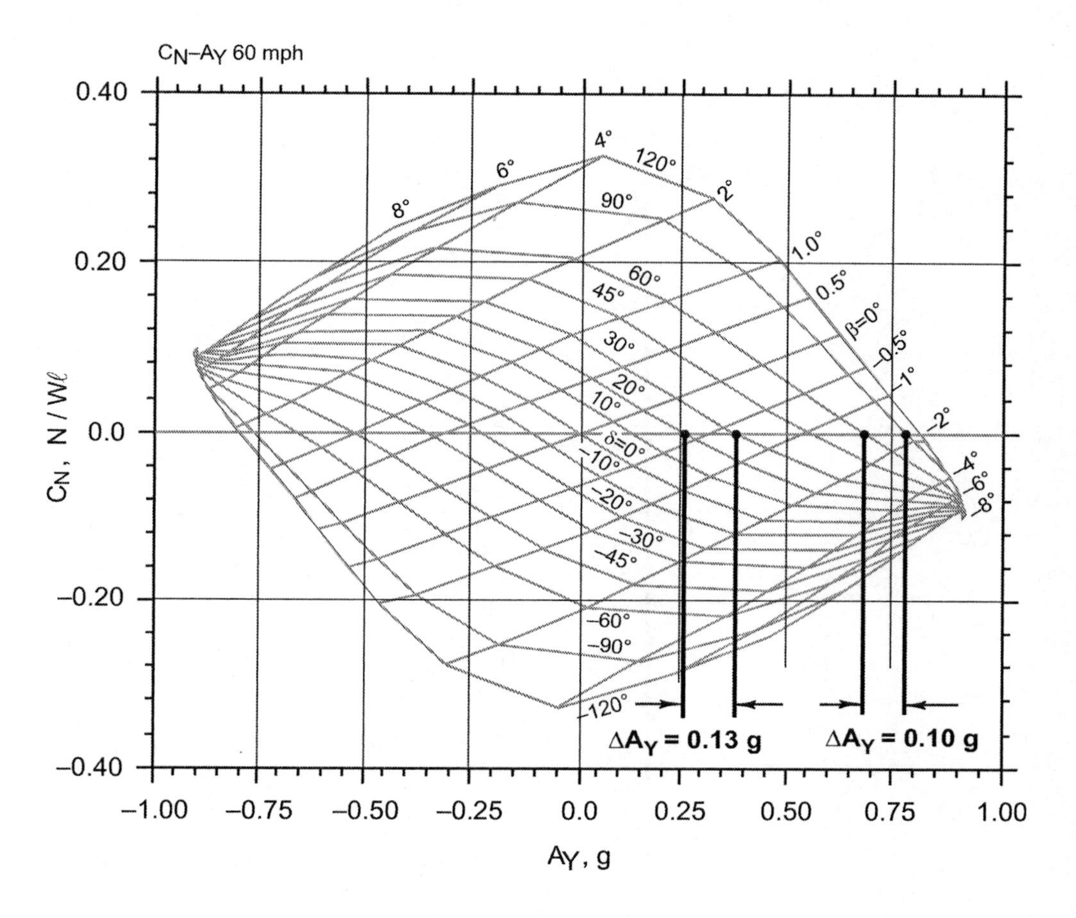

Figure 8.8 *Steering sensitivity*

Figure 8.9 shows how to measure sideslip sensitivity on a C_N-A_Y diagram.

At $A_Y = 0.3\text{g}$, , sideslip sensitivity $= -0.5/0.31 = -1.61$ deg./g

At $A_Y = 0.7\text{g}$, , sideslip sensitivity $= -1.0/0.30 = -3.33$ deg./g

e. On the C_N-A_Y plot, the understeer gradient is obtained by inverting the steering sensitivity, correcting to the front wheel angle and subtracting the Ackermann Gradient. Thus:

$$\text{UG} = \left(\frac{d\delta_{SW}}{dA_Y}\right)\left(\frac{1}{G}\right) - \left(\frac{\delta_{Acker}}{dA_Y}\right) = \left(\frac{d\delta_{SW}}{dA_Y}\right)\left(\frac{1}{G}\right) - \frac{858\ell}{V^2}$$

G = Overall steering ratio = 20 to 1
ℓ = Wheelbase = 8.5 ft.

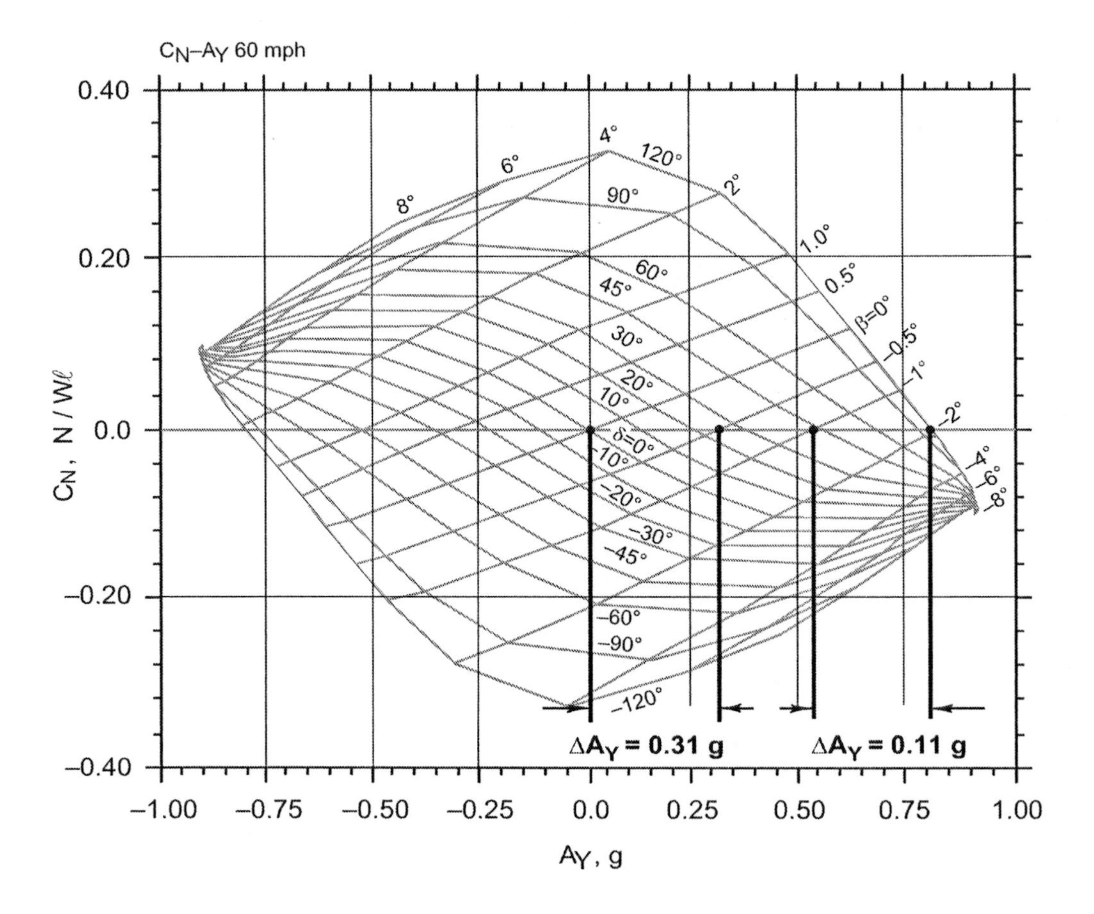

Figure 8.9 *Sideslip sensitivity*

V = Velocity = 60 mph
At $A_Y = 0.3g$:

$$UG = \left(\frac{100}{1.18}\right)\left(\frac{1}{20}\right) - \frac{(858)(8.5)}{60^2} = 4.237 - 2.026 = 2.211 \text{ deg./g}$$

At $A_Y = 0.7g$:

$$UG = \left(\frac{100}{0.37}\right)\left(\frac{1}{20}\right) - 2.026 = 13.514 - 2.026 = 11.49 \text{ deg./g}$$

f. Figure 8.10 is a plot of steering wheel angle vs. lateral acceleration along the trim line. Data for this curve was taken from Figure 8.3. The Ackermann Gradient is also shown.

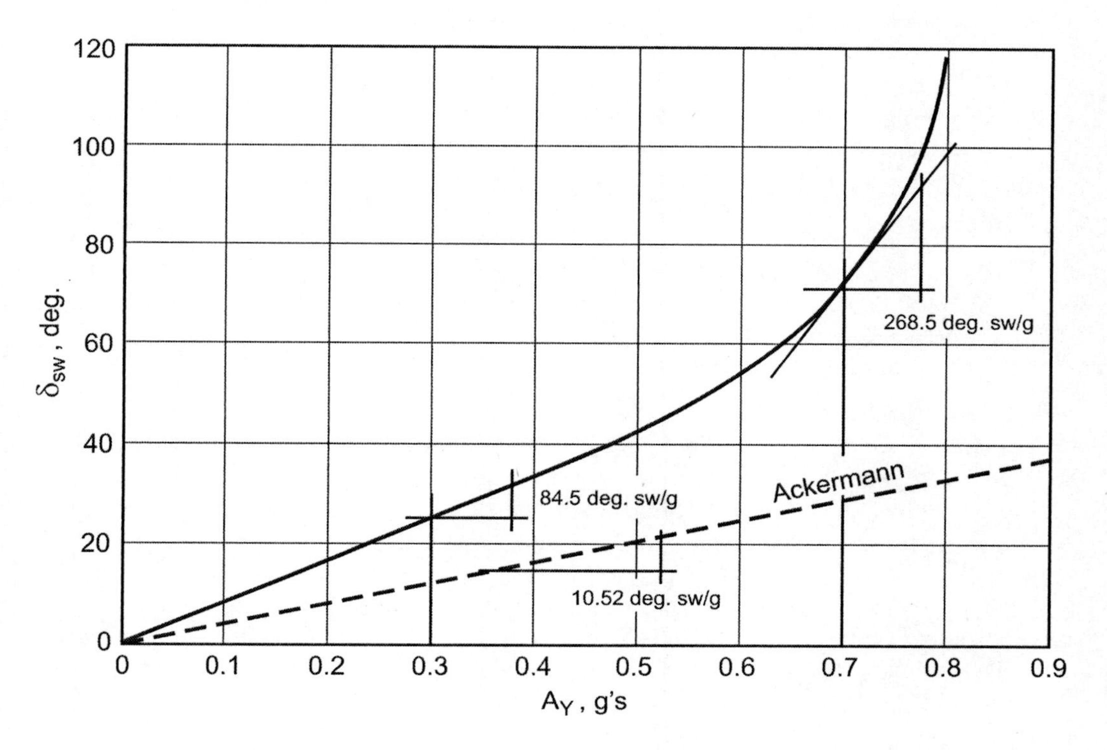

Figure 8.10 *Constant speed steady-state response test*

Slopes are measured off the figure and divided by the steering ratio (20:1). A comparison with the calculations of Part 3e are given below:

A_Y g	Calculated in 3e deg./g	Measured on Figure 8.10 deg./g
0.3	2.211	2.2
0.7	11.49	11.4

g. In the low lateral acceleration range ($A_Y < 0.3g$), the longitudinal CG location and the tire cornering slopes (lateral force versus slip angle) are major determinants of the understeer gradient, which is linear in this range (for this vehicle).

At high lateral accelerations, the large slope change in understeer gradient is primarily determined by the decrease in steering sensitivity as the cornering force of the front tires falls off. This effect is an artifact associated with using the steer angle as a measure of the yawing moment (valid only in the linear range). Said another way, the difference is mainly due to the reduction in control effectiveness at high lateral acceleration.

This does not imply that the vehicle is becoming less stable at high lateral accelerations. The ratio of the static stability measured at $A_Y = 0.3g$ and $A_Y = 0.7g$ as calculated in 3a is $-0.40/-0.25 = 1.60$. The ratio of understeer gradients at these same accelerations as calculated in 3e gives $11.4/2.2 = 5.18$.

Thus, understeer is best thought of as a useful concept in the low lateral acceleration range. The stability index is a better measure of stability in the high lateral acceleration range and a better numerical measure of stability over the full range of lateral acceleration.

4. When traveling straight ahead the vehicle is at Point A on Figure 8.11, corresponding to $(\delta, \beta) = (0\text{ deg.}, 0\text{ deg.})$. The ramp steer results in a 45 deg. steering wheel angle. Assume the input happens rapidly enough that there is insufficient time for vehicle sideslip to build while the ramp is in progress. In other words, assume β is constant during the ramp steer. Thus, the vehicle state moves along a constant β-line to Point B, $(\delta, \beta) = (45\text{ deg.}, 0\text{ deg.})$.

   The steer input is now held at 45 deg. until steady-state is reached. That is, until the vehicle sideslip reaches a value which brings the vehicle into yaw moment trim. The vehicle state progresses down a constant δ-line to the $N = 0$ axis. This is shown as Point C, corresponding to $(\delta, \beta) = (45\text{ deg.}, -1\text{ deg.})$.

   The process repeats for the removal of the steer input. Assume β does not change while the ramp steer is in progress. The vehicle state moves along a constant β-line to Point D, $(\delta, \beta) = (0\text{ deg.}, -1\text{ deg.})$. Then, the vehicle seeks equilibrium along a constant δ-line, returning to Point A, $(\delta, \beta) = (0\text{ deg.}, 0\text{ deg.})$.

   In reality, the trace is not this simple, although the basic concepts remain the same. Figure 8.12 shows the trace for the first half of this maneuver produced by MRA's VDMS dynamic simulation. The ramp steer from 0 deg. to 45 deg. occurs between A and B. Vehicle sideslip begins to build before the ramp steer is complete—although the trace near Point A begins along a constant β-line it does not follow this line all the way to B. Between B and C the steering wheel is held at 45 deg. and the vehicle searches for steady-state. The spiral approach to Point C illustrates both the coupling between the lateral and yaw response as well as the underdamped nature of the vehicle in yaw and sideslip.

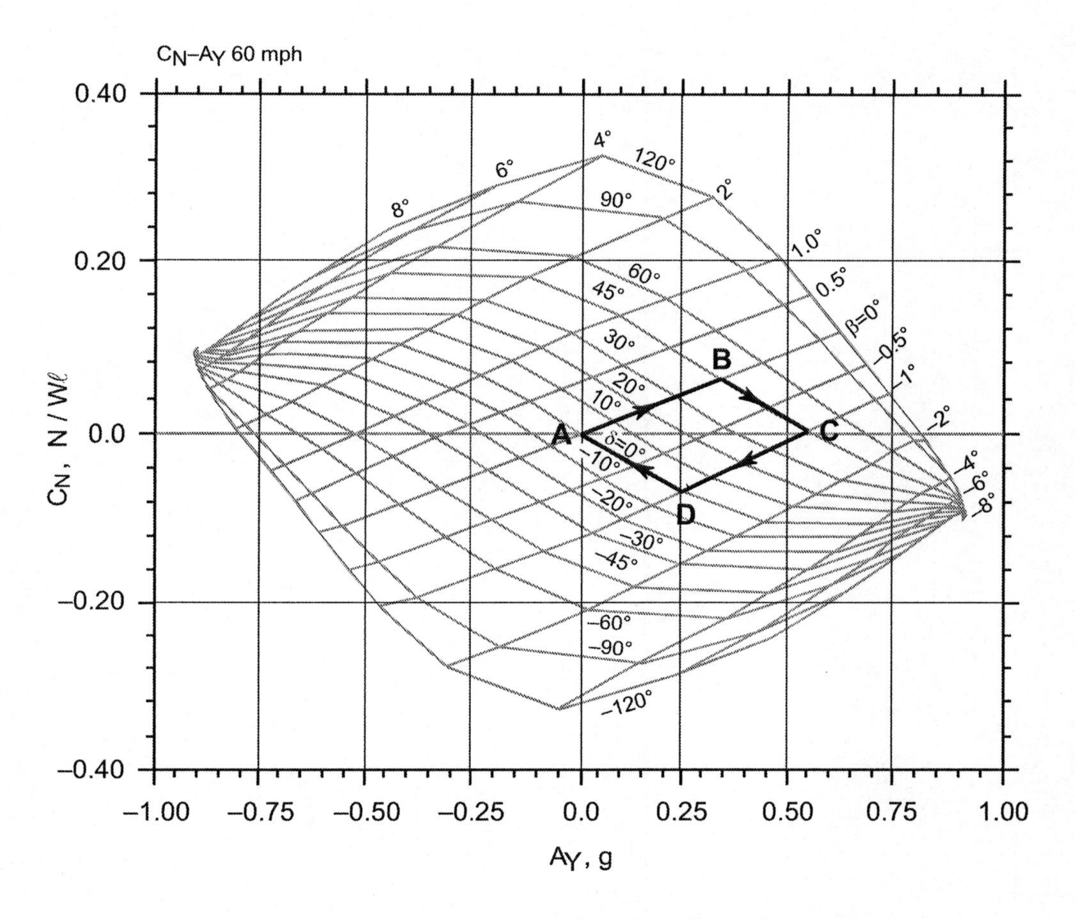

Figure 8.11 *Transient traced on the MMM diagram*

5. a. The solution to Problem 9 in Chapter 6 yields the following formula for yaw moment in the presence of a linearly increasing lateral acceleration $A_Y = ct$, where c is constant: $N = I_Z c/V$.

The lateral acceleration builds from 0 to 3.3g over a 600 ft. distance. The speed on this interval is constant at 204.5 mph or 300 ft./sec. The 600 ft. are traversed in 2 sec., leading to $c = 3.3/2 = 1.65$ g/sec.

$$N = \frac{I_Z c}{V} = \frac{\left(503.1 \text{ slug-ft.}^2\right)(1.65 \text{ g/sec.})\left(32.2 \text{ ft./sec.}^2\right)}{300 \text{ ft./sec.}} = 89 \text{ lb.-ft.}$$

$$C_N = \frac{89}{14400} = 0.0061$$

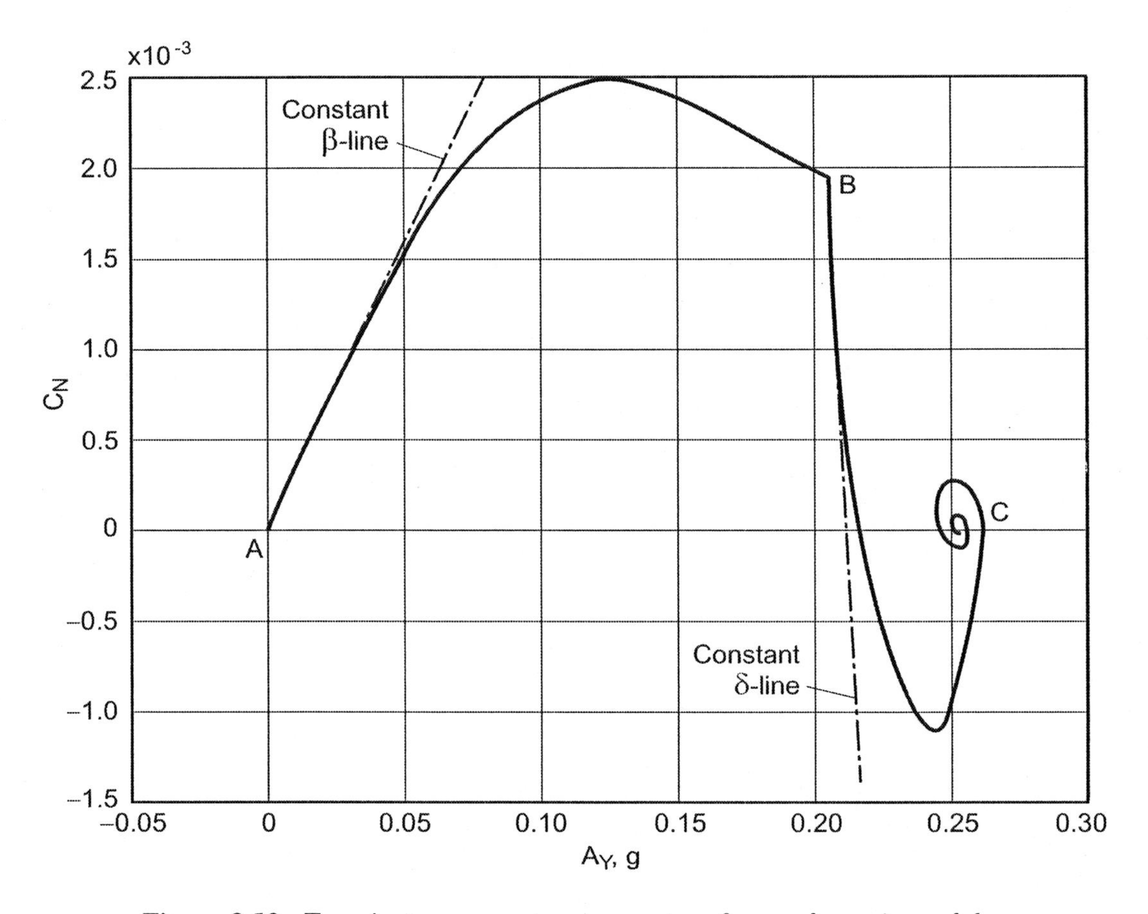

Figure 8.12 *Transient response to a ramp steer from a dynamic model*

b. The new moment of inertia is calculated from:

$$I_z = mk^2 = \left(\frac{1800}{32.2}\right)4^2 = 894.4 \text{ slug-ft.}^2$$

This gives:

$$N = \frac{I_z c}{V} = \frac{\left(894.4 \text{ slug-ft.}^2\right)(1.65 \text{ g/sec.})\left(32.2 \text{ ft./sec.}^2\right)}{300 \text{ ft./sec.}} = 158.4 \text{ lb.-ft.}$$

$$C_N = \frac{158.4}{14400} = 0.011$$

The significant increase in yaw inertia results in an increase in yaw moment and yaw moment coefficient, but the values are still very small. It would not be very noticeable to the driver.

c. The yawing moment coefficients developed in 5a and 5a are roughly 10% the size of the typical coefficient from 1 deg. of front wheel steer for this type of race car. This indicates that the driver's entry to Turn One at Indianapolis is *very* smooth. The vehicle is kept near $C_N = 0$ throughout turn entry and does not approach its ultimate capability to produce yaw acceleration.

   Because of this, yaw inertia is not of primary importance in designing a car for superspeedways. Packaging is normally directed at obtaining the desired longitudinal CG location and minimizing CG travel with fuel burn-off. The goal of minimizing yaw inertia becomes more important when designing a road course car. For rally cars, where very high angular accelerations are routinely achieved, reducing or controlling yaw inertia in the design process is a principal design goal. See pages 312-313 of RCVD for a further discussion.

CHAPTER 9

"g-g" Diagram

Solutions to this chapter's problems begin on page 124.

9.1 Problems

1. Consider the "g-g" diagrams in RCVD Figure 1.5, p.10, generated by Mario Andretti (a), Jackie Stewart (b) and Ronnie Peterson (c) while driving for Lotus at Circuit Paul Ricard, 1977. Comment on their different driving styles.

2. Since the "g-g" diagram is a function of vehicle speed, it is possible to visualize a three-dimensional diagram with longitudinal and lateral accelerations plotted on the x and y axes, and speed plotted on the z axis. Sketch such a three-dimensional figure for a sedan racing car with little/modest downforce and for an F1 car with large aero downforce.

3. Explain how each of the following increases the size of a car's "g-g" diagram:

    a. Additional aerodynamic downforce

    b. Higher friction coefficient tires

    c. Larger brakes

d. Addition of a turbocharger or supercharger

e. Lowering the center of gravity

f. Repaving of a race circuit

4. [1]Sketch the approximate shape of a "g-g" diagram boundary for each of the following vehicles. Assume dry pavement unless otherwise stated.

   a. A typical passenger car at low speed

   b. A typical passenger car at low speed driving on snow

   c. A Formula 1 race car at 100 mph

   d. A NASCAR sedan at mid-corner speed on a cornering limited oval

   e. A NASCAR sedan at top speed on a power limited oval

   f. A drag racer

5. The MRA Lap Time Simulation program was used to run two configurations of a simulated CART car on a single interesting corner of a road course. The two configurations correspond to:

   - Zero aerodynamic downforce
   - Medium aerodynamic downforce from the underbody as well as front and rear wings

   The velocity trace for each configuration is given in Figure 9.1. Both vehicles followed the same "line" and the path curvature along this segment of track is given in Figure 9.2.

   Figure 9.2 plots curvature × 1000. To determine the local path radius, R, divide by 1000 and invert. For example, a path curvature of 3.6 gives R = 1000/3.6 = 278 ft.

   Note that the "turn" that occurs between 8800 and 9700 ft. on the path distance scale has two pronounced changes of curvature, one at the beginning and the other at the end. This is a double-apex right hand turn.

   Divide the given curves into about 10 segments. Extract values from the curves at these points to calculate the longitudinal and lateral accelerations at each station for both the zero and medium downforce configurations.

[1]This problem was inspired by the *g·Analyst Learning Guide*, Valentine Research Inc., 1986.

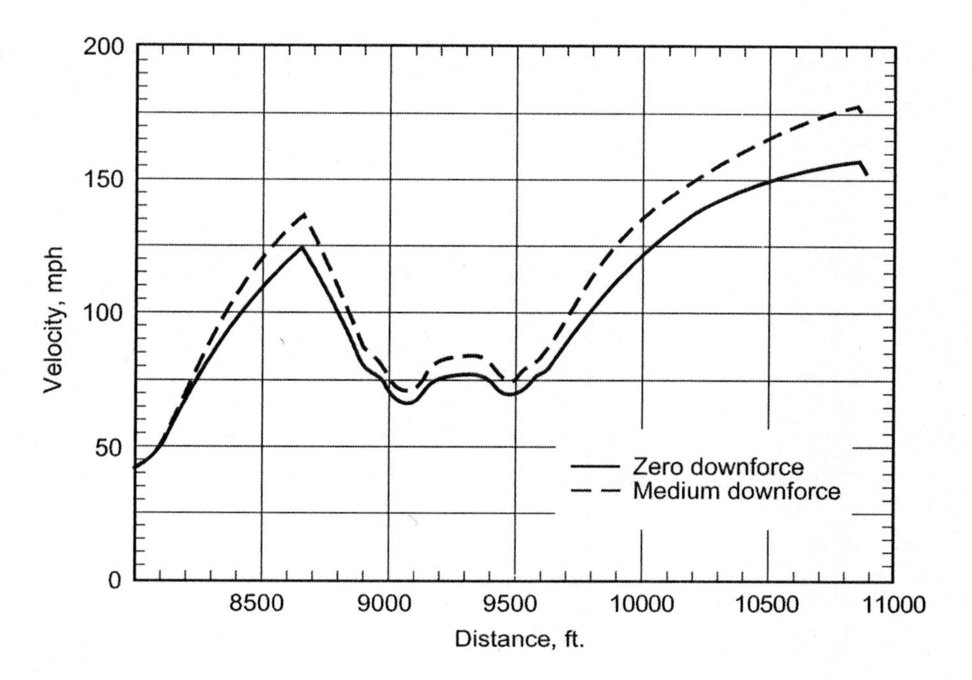

Figure 9.1 *Velocity trace with zero and medium downforce*

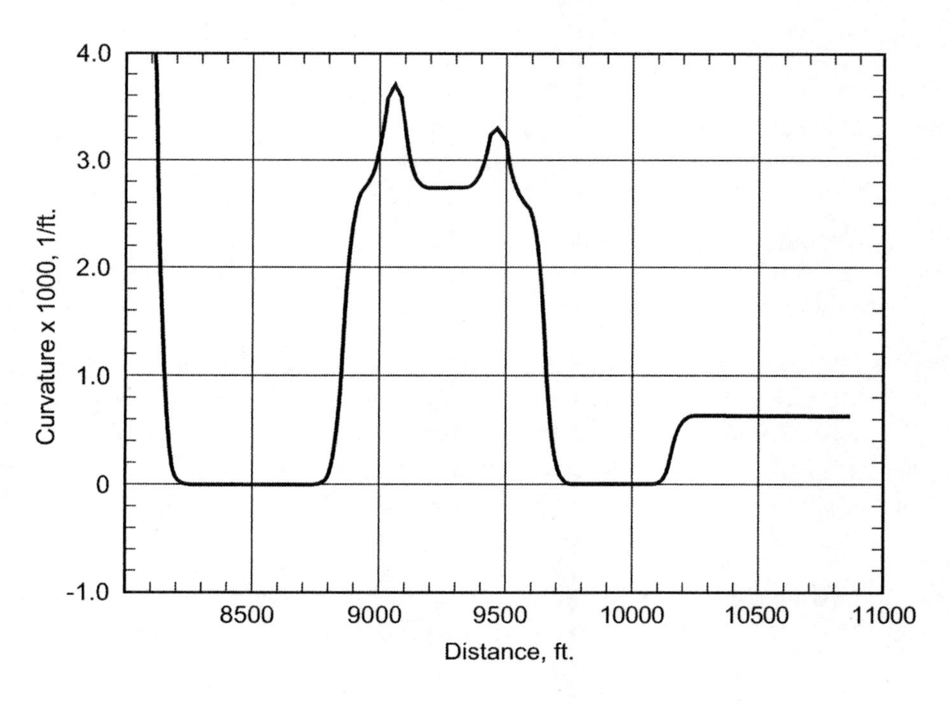

Figure 9.2 *Path curvature*

Plot these acceleration components on a "g-g" diagram, where the scales have the same sense as RCVD Figure 1.5, p.10. Since aero downforce is a function of speed, the plotted points may diverge from a single smooth elliptical shape as in RCVD Figures 9.10 and 9.11, pp.358-9.

9.2 Simple Measurements and Experiments

NOTE: All experiments should be carried out legally, at low speeds (say, below 30 mph) and under safe conditions. Keep in mind that it is possible to roll-over *any* vehicle, given the proper circumstances. Experiments should be conducted in a large, paved area free of obstacles, traffic and pedestrians. You are responsible for maneuvering safely. Use good judgment in the following exercises.

The experiments in this chapter are adapted from the *g·Analyst Learning Guide*[2]. They can be used with any accelerometer, not just the g·Analyst. The exercises start by familiarizing the reader with an accelerometer's use and culminate with the measurement of your vehicle's "g-g" diagram. The exercises assume an accelerometer has been mounted in the vehicle and calibrated.

The g·Analyst features a real-time display of the acceleration plotted on the "g-g" diagram axes. A sample of the display is given in Figure 9.3.

1. This experiment concentrates on longitudinal acceleration. Find a paved section of road that's free of traffic or an empty parking lot. You will need a place where you can drive and safely watch the display.

   At rest the data dot is centered at the origin. Obviously the car is not accelerating. Now accelerate to 25 mph in a straight line and hold that speed. Notice that the data dot moved forward (up the A_X axis) then returned to the origin as you reached constant speed. This is to remind you of the difference between acceleration and speed. Acceleration is the change in speed, not speed itself. Remember this distinction, as it will be critical to your understanding of car handling.

   While moving forward at 25 mph, apply the brakes moderately and come to a full stop. Notice that the data dot moved downward (along the negative A_X axis) as you were braking and then centered itself at the origin as you came to rest.

[2]Used with the permission of Valentine Research, Inc.

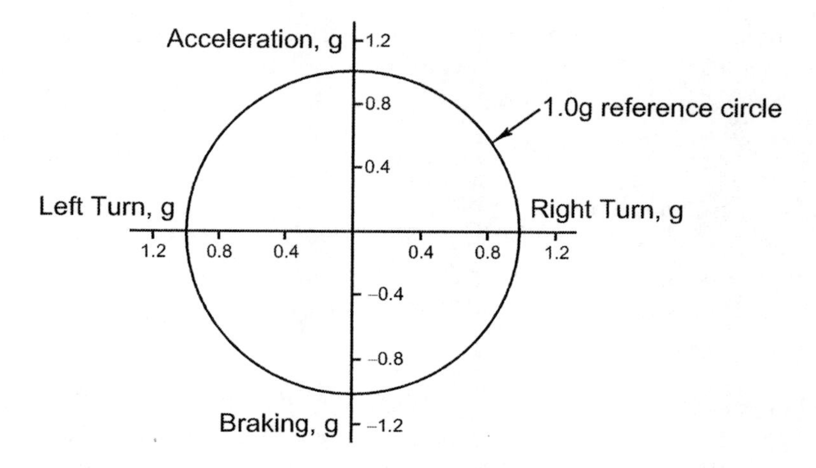

Figure 9.3 *Accelerometer display*

Accelerate briefly in reverse. Again the data dot moves downward along the negative A_X axis. This should remind you once again of the difference between speed and acceleration. It should also confirm that forward braking and backward acceleration are changes in speed of a similar nature and the accelerometer records them similarly.

Now, while driving straight ahead, experiment with harder forward acceleration and harder braking, just to get a feel for how they are represented on the display.

2. In this exercise the lateral acceleration will be studied by driving on circles of different radii at various speeds. Start by circling to the right. Hold a steady steering wheel position at a steady, moderate speed. The data dot moves to the right because the vehicle is accelerating to the right. How far it moves indicates how hard you are cornering. At a constant speed on a constant radius circle the dot remains in a constant position because the lateral acceleration is constant.

   Now increase speed while maintaining the same circle. You may have to change your steering wheel angle to maintain a constant radius. Notice that the data dot moves farther to the right as speed increases. Returning to the original speed on the same radius will cause the data dot to return to its former position.

   Now, while holding a constant speed, steer a tighter circle. The data dot moves farther to the right as the radius decreases.

   Repeating these maneuvers while steering to the left will produce similar results, except that the data dot will move to the left.

The point of the exercise is that lateral acceleration does not depend on speed alone or on tightness of the turn alone. Whenever a vehicle has a speed along a curved path it has lateral acceleration, even at constant speed. This lateral acceleration is measured by the accelerometer.

3. This experiment examines combined lateral and longitudinal accelerations. Begin with a constant speed, constant radius circle. The data dot lies on the axis to the right (for a right hand turn), indicating pure lateral acceleration. Apply power to increase speed and watch the data dot move above the horizontal axis. This indicates both lateral and longitudinal acceleration.

   While on the same circle reduce speed. The dot drops below the horizontal axis, indicating a combination of lateral and longitudinal (this time rearward) acceleration. Continue to slow until you stop. The data dot returns to the origin, indicating no accelerations.

   Whenever the data dot is displaced from the origin the vehicle is experiencing an acceleration. The magnitude of the acceleration is given by the straight line distance from the origin to the data dot.

4. This exercise will determine the maximum longitudinal acceleration capability of your vehicle. In this and subsequent experiments use your accelerometer's recording capability so that you can concentrate on driving the maneuvers. Play back the recorded data afterward to extract the desired information. Visually reading data off the display while driving the maneuver is discouraged.

   To determine the maximum forward acceleration, drive in a straight line at walking speed and apply full power. The data dot will move forward and a g-value will be displayed. The value depends on the weight and engine output of the vehicle. Note the maximum forward acceleration you can achieve.

   Acceleration will diminish as speed increases because of gear selection and because of increasing aerodynamic resistance. If your vehicle is powerful enough to spin its wheels when accelerating from rest, then wheelspin will limit forward acceleration at low speeds. At higher speeds forward acceleration will be power limited.

   Now drive in a straight line at a modest speed and apply the brakes as hard as possible without locking the wheels. Work up to maximum braking with repeatedly severe stops, being careful not to lock a wheel. Lock-up produces very rapid tire wear and perhaps even flat spots. Note the maximum g-value you can achieve while braking. Typically, maximum deceleration is achieved without locking wheels.

5. This experiment determines the maximum lateral acceleration capability of your vehicle. Before starting this experiment, note that prolonged driving at high lateral accelerations can cause the oil pump pickup to uncover, resulting in a loss of oil pressure. Substantial tire wear can also result.

   Mark a constant radius circle with traffic cones or with chalk. A 50 ft. radius is sufficient. Drive clockwise around the circle at a slowly increasing speed. As speed increases you will either have to increase the steering wheel angle (if the vehicle is understeer) or decrease the steering wheel angle (for oversteer) to stay on the same radius. Continue increasing speed until you cannot stay on the circle or until full power is reached. Review the accelerometer data to determine the maximum lateral g-value.

   Approach the limit gradually as vehicle response near the limit is very different from that in ordinary passenger car driving. Make sure there is ample run-off room in case something unexpected happens.

6. In this exercise, the full "g-g" diagram is sought. The g·Analyst has a display option called "signature mode". In this mode, the acceleration dot leaves a trail, similar to the "g-g" diagram in RCVD Figure 9.7, p.355. The idea is to conduct a number of individual maneuvers that will fill-in the complete "g-g" diagram and, in doing so, define the boundary as shown in Figure 9.4.

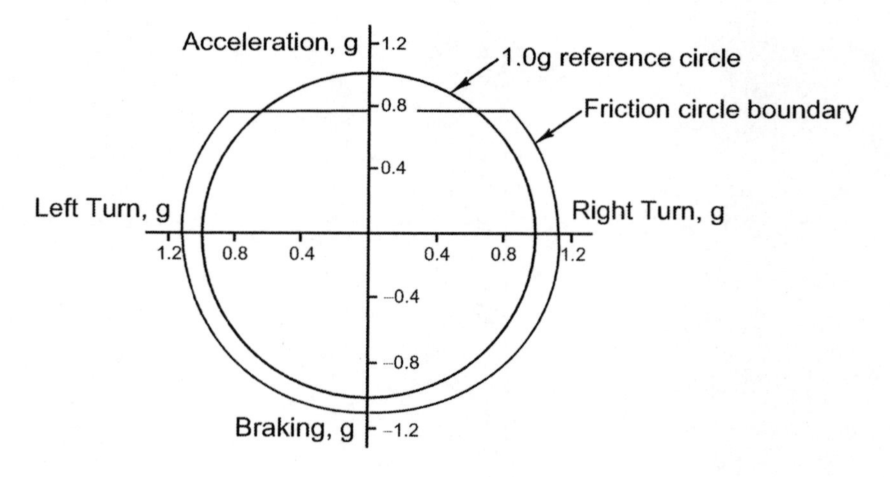

Figure 9.4 *Accelerometer display*

To fill-in the area in the first quadrant, accelerate at full power from a low speed in a low gear while steering to the right. Repeat at increasingly tight steer angles. Repeat the full cycle of steering inputs at full power in a higher gear, or use less than full power.

To fill the second quadrant, conduct the same series of runs while steering left instead of right. For quadrants three and four, begin each maneuver at a moderate speed and apply the brakes as you turn. Apply different amounts of brakes and, at each braking level, input increasingly tight steer angles. Do this for both left and right hand turns.

Run as many maneuvers as necessary to produce a "g-g" diagram with smooth boundaries. The result is a "g-g" diagram for your vehicle.

7. Make a very simple course in your test area and drive a lap. Look at the accelerometer trace and identify each of the maneuvers you performed.

9.3 Problem Answers

1. Andretti's "g-g" diagram (a) extends much farther than the other two drivers in negative longitudinal acceleration. It indicates heavy braking, and probably very late braking. Note how he is avoiding more than 1g of lateral acceleration in left hand turns while braking above -1g. Andretti's diagram also has very few points near the center of the diagram, meaning that he is smoothly transitioning from braking to cornering to accelerating at each turn.

   Stewart's diagram (b) appears to be smaller than the other two. This does not necessarily indicate that he is slower, but rather that he is consistently driving just below the vehicle's limit. It indicates a very smooth and precise driver. He is probably devoting a large part of his driving effort to placing the car on the best possible line around the circuit. Stewart had also retired the previous year, so he was probably not "hanging it out" as much as he might have in previous years.

   Peterson, on the other hand, has a wider and more chaotic "g-g" diagram (c). Data points are clustered around the edges of the diagram, with numerous "outliers". This diagram indicates a very aggressive driving style where the vehicle is kept at its lateral limit nearly all the time.

2. Figure 9.5 presents the "g-g-V" diagram for a sedan racing car with little/modest downforce. Figure 9.6 is for an F1 car with large aero downforce. Notice how downforce increases the size of the diagram at moderate to high speeds.

3. a. Additional aerodynamic downforce increases the normal loads on the tires. This, in turn, results in greater lateral and longitudinal forces from the tires at a given slip angle or slip ratio. Larger accelerations result at speeds where aero is significant.

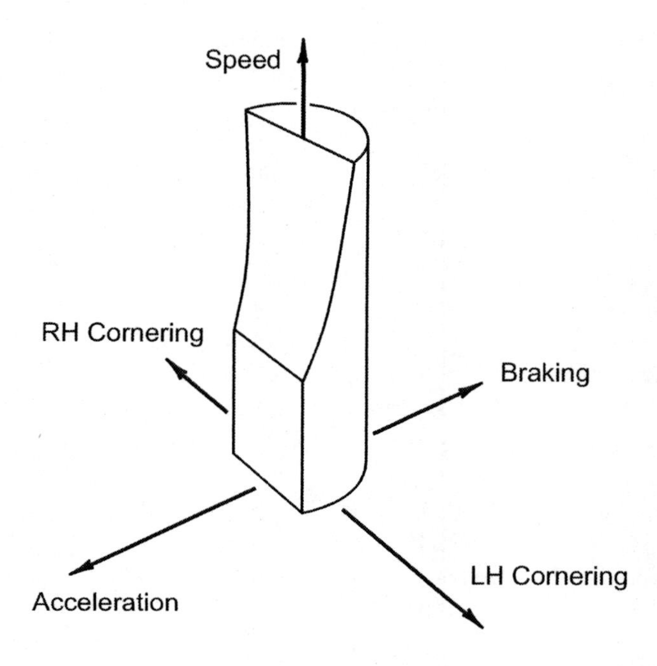

Figure 9.5 *"g-g-V" for a sedan with little downforce*

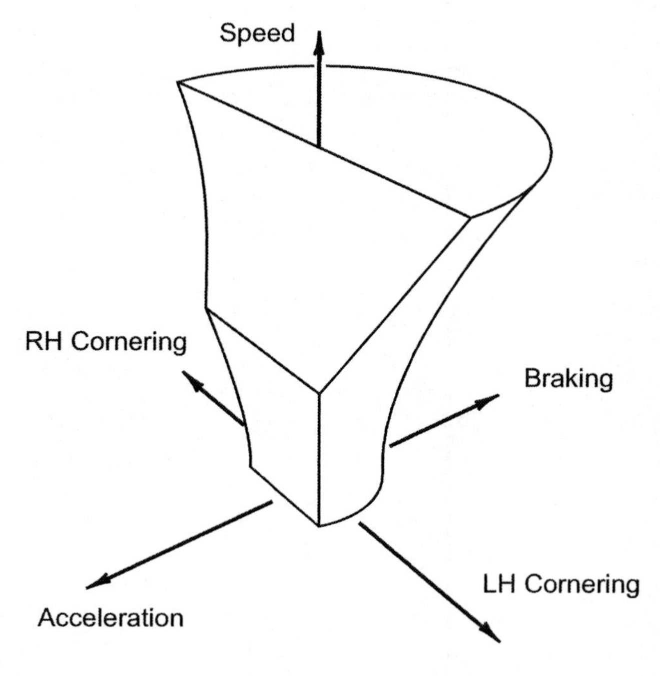

Figure 9.6 *"g-g-V" for a F1 car with ample downforce*

b. Higher friction coefficient tires increase the amount of lateral or longitudinal force a tire can produce at a given normal load. This increases the vehicle's acceleration capability in all directions.

c. Larger brakes have the ability to provide more braking torque and greater deceleration, provided the tires are capable of providing the additional longitudinal forces. If the tires were already the limiting factor (able to lock all four wheels with the original brakes) then adding larger brakes will not increase the size of the "g-g" diagram.

d. The addition of a turbocharger or supercharger should increase the power output of the engine. All "g-g" diagrams are power limited above a certain speed. In these regions additional horsepower will expand the diagram in the acceleration direction.

e. Lowering the center of gravity reduces lateral load transfer while cornering. This allows the pair of wheels across each axle to provide more lateral force because they will be operating at more equal normal loads. More lateral force implies more lateral acceleration, increasing the diagram boundary. For rear drive cars, lowering the CG decreases the longitudinal load transfer during acceleration, reducing the load on the rear tires and the size of the "g-g" diagram when traction limited. Vice versa for front drive cars.

f. Repaving of a race circuit can increase a car's "g-g" diagram, evidenced by lower lap times. New surfaces tend to be smoother than the old ones. Fewer bumps results in more even tire loads and greater lateral force generation. Smooth tracks also allow for lower ground clearances (good for aero downforce) and less suspension movement (less ride-steer and roll-steer effects). Furthermore, a new surface may allow for increased adhesion of the tires to the surface.

4. The corresponding "g-g" diagrams are given in Figure 9.7. Comments on each are given below.

a. A typical passenger car has relatively weak tires. At low speeds the tires may be limiting in forward acceleration. If so, the "g-g" diagram is nearly circular.

b. On snow the tire/road friction coefficient is greatly reduced. As a result the "g-g" diagram is very small.

c. A Formula 1 race car at 120 mph produces a considerable amount of downforce. This, combined with its powerful engine, extremely large brakes, light weight and high performance tires leads to a very large "g-g" diagram boundary.

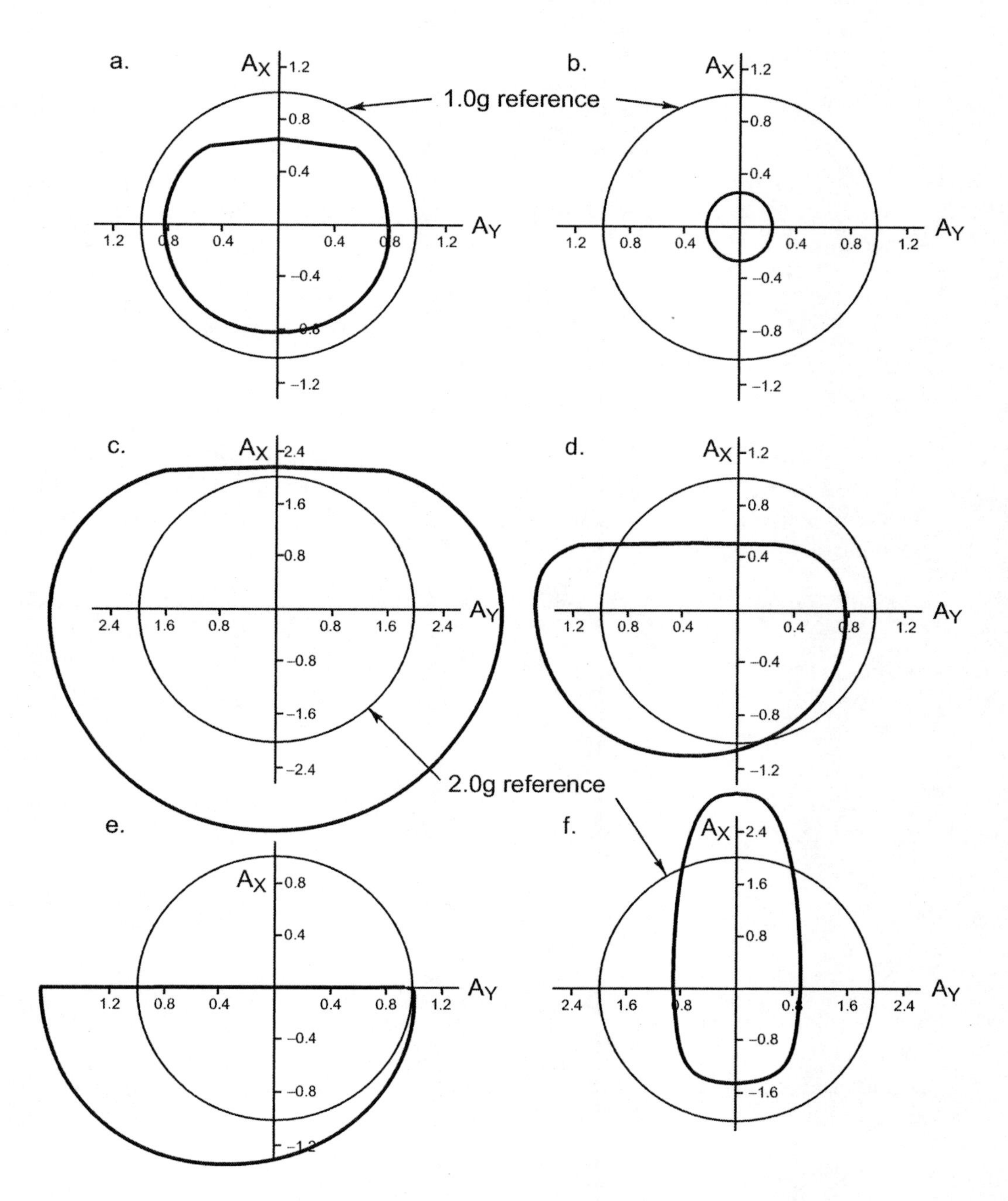

Figure 9.7 *"g-g" diagrams for various vehicles*

d. Since NASCAR sedans only turn left on oval tracks the asymmetric set-up skews the "g-g" diagram. Since the vehicle is not at top speed it has some amount of forward acceleration available, although this is clearly power limited.

e. A NASCAR sedan at top speed on a power limited track such as Daytona or Talladega has no capability to increase its speed. Thus, no forward acceleration capability is indicated on the "g-g" diagram.

f. A drag race does not involve any turns, only straight-line acceleration. Thus, the cars are optimized for large forward accelerations, leading to a narrow, elongated "g-g" diagram. Note that the addition of deployed parachutes would extend the diagram in the braking direction, but the effect would diminish as speed decreased.

5. Figure 9.8 shows the lateral and longitudinal accelerations calculated on the corner of interest for the zero downforce case as well as speed segregated "g-g" diagrams for a complete lap of this circuit. Note how, in the zero downforce case, the boundaries are approximately independent of speed.

   Similarly, Figure 9.9 shows the lateral and longitudinal accelerations calculated on the corner of interest for the medium downforce case as well as speed segregated "g-g" diagrams for a complete lap of this circuit. Note how, in the presence of aerodynamic downforce, the boundaries of the "g-g" diagram enlarge with increasing speed.

9.4 Comments on Simple Measurements and Experiments

As with the experiments themselves, the comments below are adapted from the *g·Analyst Learning Guide*[3].

1. When braking, an occupant's head wants to tilt forward in response to a force. Therefore, a common first response is to think of braking force and, for that same reason, accelerating and cornering *forces*. Why, then, measure acceleration? Wouldn't some measure of force be better?

   The answer is both "yes" and "no". In vehicle handling our concern is managing the motions of a car through maneuvers such as cornering or braking. Acceleration is a motion variable and can be *directly measured*. Forces are more fundamental,

[3] Used with the permission of Valentine Research, Inc.

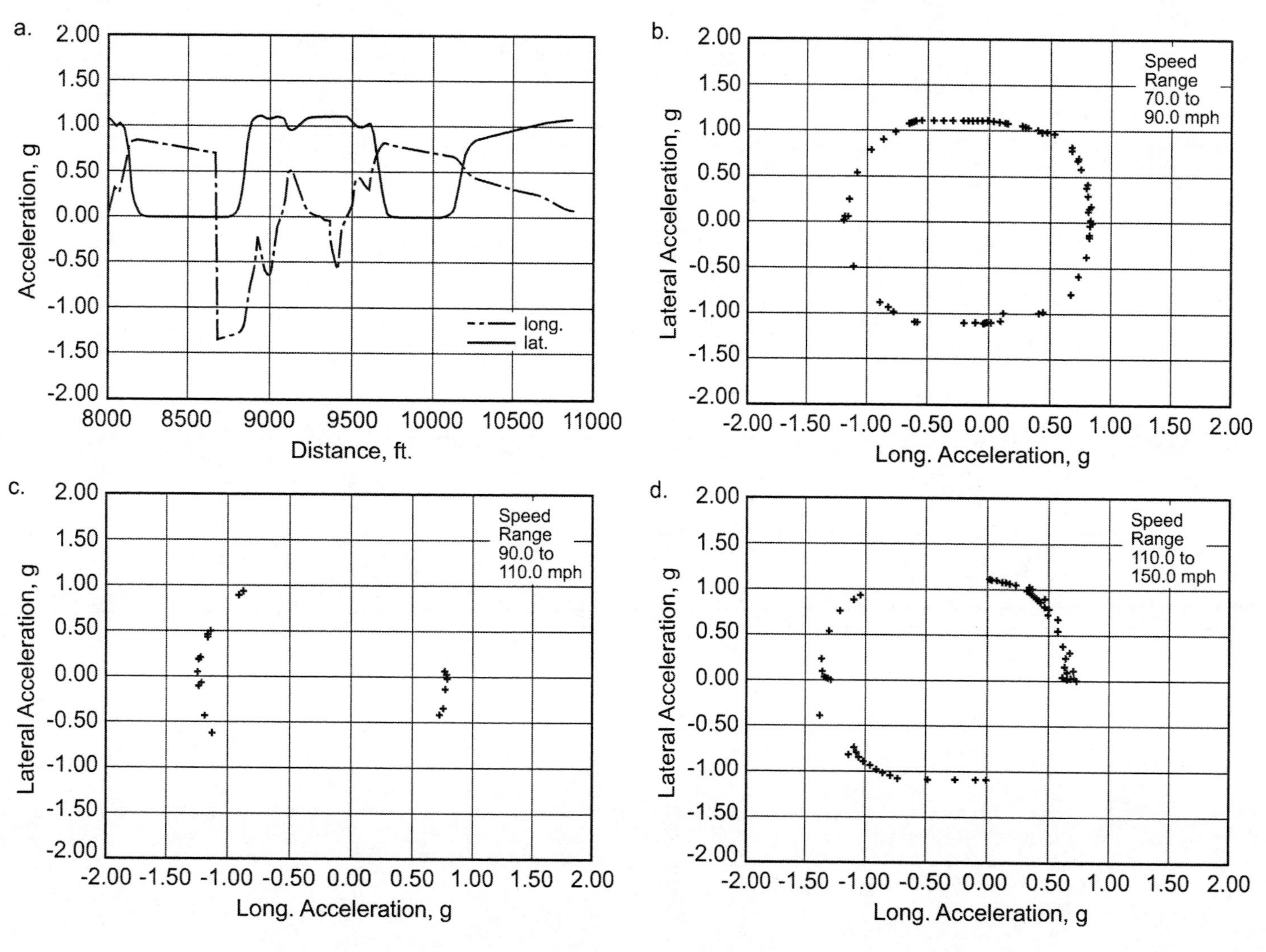

Figure 9.8 *Zero downforce case*

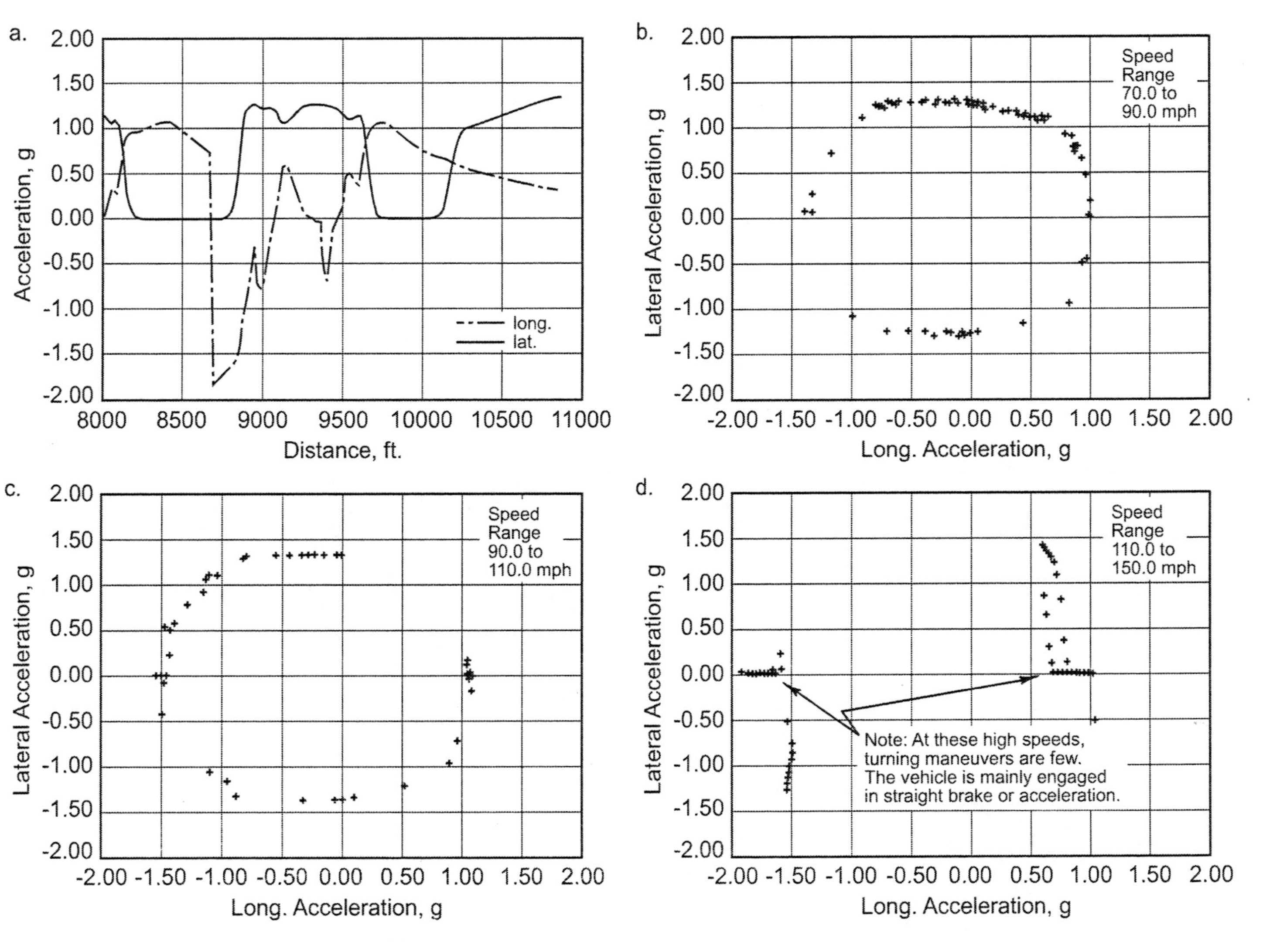

***Figure 9.9** Medium downforce case*

but are extremely difficult to measure. Since the mass is never exactly known, both the forces and the mass are secondary to the acceleration which, in racing, is to be maximized at all times.

Accelerations, however, are filtered by the inertia (mass) of the vehicle. Analysis of the vehicle using force-moment techniques provide many useful insights (see RCVD Chapter 8).

2. The effect of changes in speed (V) or radius (R) on lateral acceleration (A_Y) are given by the equation $A_Y = V^2/Rg$. The equation says that doubling the speed on a given corner will quadruple the resulting lateral acceleration, while maintaining the same speed on a corner of half the radius will only double the lateral acceleration. Cornering is much more sensitive to speed than radius, although both contribute to lateral acceleration. Given a speed-independent "g-g" diagram the maximum speed for a corner will vary in proportion to the square root of the radius.

3. The location of the data dot indicates the combined lateral and longitudinal acceleration. These components add vectorally to the total acceleration. Depending on the type of accelerometer, the output may be a single, combined acceleration or the individual components may be specified. The g·Analyst offers a choice of displays so that output may be viewed in either form. If your accelerometer outputs the components or outputs only the total, the "missing" output can be determined using simple vector calculations.

4. Typically, one wheel will lock before (sometimes well before) the others. This can induce a yaw moment on the vehicle and, without steering correction, path curvature. Further increasing pedal pressure will typically cause a pair of wheels on one end of the vehicle to lock. If the front wheels lock first the ability to steer is lost and the vehicle tracks more-or-less straight ahead. If the rear wheels lock first the vehicle loses directional stability and may spin, even in a straight-line braking maneuver. All four wheels rarely lock simultaneously.

5. With maximum lateral acceleration now measured, compare against the maximum braking and forward acceleration values. Typically, a road car will produce somewhat smaller maximum lateral acceleration values than braking values. Due to vehicle asymmetries it is not uncommon for the left and right hand maximum lateral accelerations to differ. The maximum forward acceleration will be noticeably smaller than the other values whenever the forward acceleration is power limited.

6. A sequence of individual maneuvers are suggested to find the "g-g" diagram boundary instead of a single maneuver. The boundaries could be found from a lap of a suitably designed race circuit (as in RCVD Figure 9.7, p.355), but this involves

much more skill. Also, there is no guarantee that a given race circuit will exercise the full "g-g" boundary of a vehicle.

7. Once familiar with the workings of the accelerometer you can use it to analyze driver/vehicle performance at specific points around a circuit. You may be able to identify regions where the vehicle needs to be improved (set-up or design changes), or parts of the circuit where the driver is not operating the vehicle at its limit.

CHAPTER 10

Race Car Design

Solutions to this chapter's problems begin on page 136.

The problems in this chapter are focused on a simple preliminary design exercise for a three-wheeled vehicle. Two wheels are in the front (front drive) and one on the rear. This vehicle is referred to as the "**CH10**" and has the following parameters:

- Wheelbase = 100 in.
- Gross weight = 2000 lb.
- Sprung mass CG height = 15 in. above ground
- Unsprung weight = 50 lb./wheel
- Front track = 60 in.
- Roll axis height = 0 in. (on the ground)
- Ride spring linkage ratio = 1.0 in./in.
- Anti-roll bar linkage ratio = 1.0 deg./deg.
- Ackermann: Parallel
- Tire vertical spring rate = 2000 lb./in. (assume constant)
- Tires: P275/40ZR17 Eagle ZR, Figure 10.1.

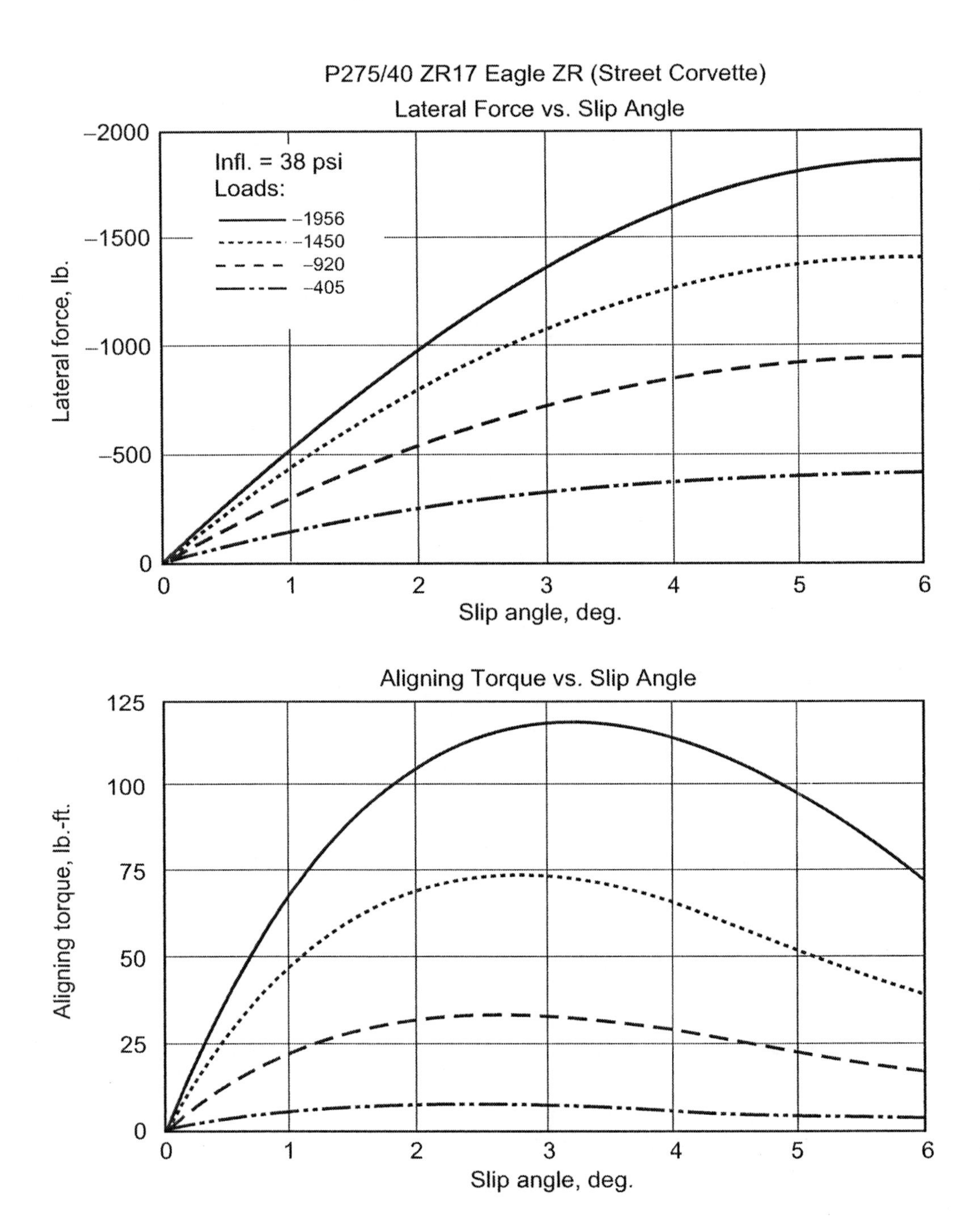

Figure 10.1 *Tire data for use with the CH10.*

10.1 Problems

1. Determine the horizontal CG location required to give the CH10 linear range neutral steer. For this calculation, assume no lateral load transfer due to cornering. Cross plot and interpolate the tire data as necessary. Assume free rolling conditions exist, i.e., no tractive-effort effects on the tire data.

2. Using the CG location calculated in Problem 1, determine the overall ride rates (chassis to ground) and wheel center rates (chassis to wheel center, equal to spring rate on this car) for a ride natural frequency of 90 cycles per minute.

3. Using the results of Problems 1 and 2, calculate the anti-roll bar rate (on the front only, of course!) to give a roll gradient of 1.5 deg./g. How much of the total roll rate is provided by the ride springs and how much by the anti-roll bar?

4. Assuming the CG location from Problem 1, will the anti-roll bar from Problem 3 give terminal push, terminal spin or terminal drift at the limit? If the vehicle is not balanced, calculate a new longitudinal CG location to make the vehicle terminal drift at the limit.

5. Using the CG location you determined in Problem 1, calculate a brake distribution that will give front wheel lock-up first in 1g straight line braking.

6. Sketch the side view geometry of an independent front suspension to give 50% anti-dive (outboard brakes). Assume a front brake percentage of 80%.

7. Suppose the CH10 develops 500 lb. of downforce on the front and 500 lb. of downforce on the rear at a certain speed. Repeat the calculation of Problem 1 to include the effect of aero downforce. Namely, determine the horizontal CG location required to make the CH10 neutral steer in the linear range at this speed.

8. These problems become more challenging when a four-wheel vehicle is used. Generate additional input data as needed when solving the four-wheel case.

10.2 Problem Answers

1. For neutral steer, RCVD Chapter 5 states that the longitudinal CG location and neutral steer point coincide. Thus:

$$\mathrm{NSP} - \frac{a}{\ell} = 0 \quad \Rightarrow \quad \frac{C_R}{C} = \frac{a}{\ell}$$

where $C = C_{LF} + C_{RF} + C_R$. Calculating the wheel loads:

$$F_{zLF} + F_{zRF} = \frac{Wb}{2\ell}$$

$$F_{zR} = \frac{Wa}{\ell}$$

The latter of these says that:

$$\frac{F_{zR}}{W} = \frac{a}{\ell} \quad \Rightarrow \quad \frac{F_{zR}}{W} = \frac{C_R}{C} \text{ for neutral steer}$$

It follows that:

$$\frac{F_{zF}}{W} = 1 - \frac{C_F}{C} \text{ for neutral steer}$$

Now determine the cornering stiffnesses, which will be a function of load. Pulling data off of Figure 10.1 leads to the plot in Figure 10.2.

The goal is to find where the CG location and the ratio of rear cornering stiffness to the total cornering stiffness are identical. The cornering stiffness curve could be fit with a polynomial and the problem could be solved algebraically. Or, a few values could be calculated and the trend observed. The following table results from this second approach:

a/ℓ	F_{zF} one tire	C_F one tire	C_F two tires	F_{zR}	C_R	C	C_R/C
–	lb.	lb./deg.	lb./deg.	lb.	lb./deg.	lb./deg.	–
0.2	800	272	544	400	143	687	0.208
0.3	700	240	480	600	208	688	0.302
0.4	600	208	416	800	272	688	0.395
0.5	500	176	352	1000	328	680	0.482

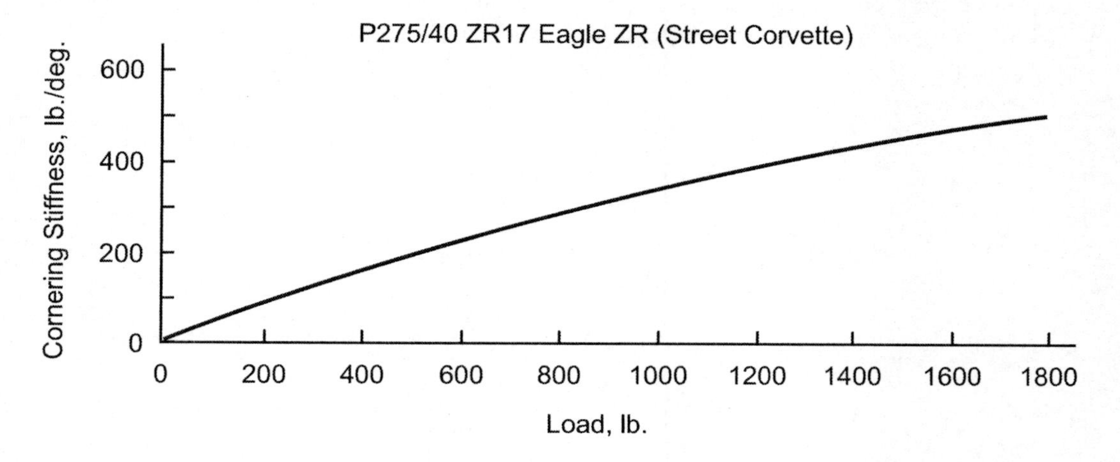

Figure 10.2 *Cornering stiffness as a function of load*

From this, a/ℓ is equal to C_R/C somewhere between 0.3 and 0.4. The answer for the linear case (obtained through careful plotting of tabular results or via a polynomial fit to the cornering stiffness curve) is $a/\ell = 0.333$. That is, the CG should be located 1/3 of the distance aft of the front axle. This result is for no lateral load transfer on the front, justified for the case of low speed, linear range neutral steer.

2. The CG is located at $a/\ell = 0.333$. This results in the following wheel loads:

$$F_{zLF} = F_{zRF} = \frac{1}{2}(1 - 0.333)\,2000 = 667 \text{ lb.}$$

$$F_{zR} = (0.333)\,2000 = 667 \text{ lb.}$$

All three wheels carry 667 lb. Therefore, the same overall spring rate K at each wheel will result in the same natural frequency for the vehicle. To arrive at a ride natural frequency of 1.5 Hz or 90 cycles per minute, the following spring rate is calculated:

$$\omega_n = \sqrt{\frac{K}{m}} \quad \Rightarrow \quad 2\pi(1.5) \text{ rad./sec.} = \sqrt{\frac{K\left(32.2 \text{ ft./sec.}^2\right)}{667 \text{ lb.}}}$$

$$K = 1840 \text{ lb./ft.} = 153.3 \text{ lb./in.}$$

Since the ride spring linkage ratio is unity, the ride spring rate is equal to the wheel center rate. The ride spring K_W is in series with the tire spring K_T. They contribute to the overall ride rate K according to the following:

$$K = \frac{K_W K_T}{K_W + K_T}$$

$$153.3 = \frac{2000 K_W}{K_W + 2000} \quad \Rightarrow \quad K_W = 165.9 \text{ lb./in.}$$

Thus, the overall ride rate is 153.3 lb./in. and the wheel center rate is 165.9 lb./in.

3. First, calculate the rolling moment per g of lateral acceleration, M_ϕ / A_Y. This is given by:

$$\frac{M_\phi}{A_Y} = hW_S = 1.25\,(2000 - (50 \times 3)) = 2312.5 \text{ lb.-ft./g}$$

Note that, with the roll axis on the ground, the moment arm about the roll axis is equal to the sprung mass CG height. Also, the three unsprung weights were subtracted from the gross weight to determine the sprung weight. RCVD p.603 relates the roll gradient RG, roll stiffness K_ϕ and rolling moment per g lateral acceleration:

$$(RG)\,K_\phi = \frac{M_\phi}{A_Y}$$

A roll gradient of 1.5 deg./g (race car stiff) is sought. Solving for K_ϕ:

$$K_\phi = \frac{1}{(RG)}\frac{M_\phi}{A_Y} = \frac{1}{1.5}(2312.5) = 1541.7 \text{ lb.-ft./deg.}$$

Ignore the roll stiffness due to the tires for the time being and assume the roll stiffness is provided by the ride springs and the anti-roll bar. Consider the contribution of one ride spring:

$$K_{\phi W} = (\text{moment arm})\,(\text{deflection})\,(\text{spring rate}) = \left(\frac{t}{2}\right)\left(\frac{t}{2}\phi\right) K$$

Converting from inches to feet, from degrees to radians and summing the contributions from both front springs leads to:

$$K_{\phi W} = \frac{Kt^2}{2 \times 57.3 \times 12} = \frac{165.9 \times (60)^2}{1375.2} = 434.3 \text{ lb.-ft./deg.}$$

The ride springs provide 434.3 lb.-ft./deg. of the required 1541.7 lb.-ft./deg. The anti-roll bar needs to supply another 1107.4 lb.-ft./deg.

These calculations ignored the tire spring rate contribution to roll gradient. The tire spring rate acts in series with the ride spring and anti-roll bar (which, themselves, act in parallel). In stiffly sprung cars the tire spring rate can have a significant effect on the roll gradient. In this example, even if the ride springs and anti-roll bar were infinitely stiff (such as in a go-kart) the CH10 would be limited to a roll gradient of 5235.6 lb.-ft./deg. The CH10 would acquire a roll angle due to sidewall deflection.

4. There are two facets to this question. The first deals with the role of an anti-roll bar on a three-wheeled vehicle. The second is concerned with the distinction between linear range and limit behavior.

   The anti-roll bar on this three-wheeler has no first-order effect on the linear understeer/oversteer behavior. It only determines the magnitude of the roll gradient. The distribution of the lateral load transfer between the front and rear tracks affects the UO balance. With this three-wheeler the single rear tire cannot provide any roll resistance and therefore does not experience any load transfer. Regardless of the size of the anti-roll bar, 100% of the roll moment will be taken on the front axle.

   Thus, the addition of an anti-roll bar to a neutral steer CH10 (CG located as per Problem 1) has no effect on the UO characteristic. The CH10 remains neutral steer in the linear range.

   As for the distinction between linear range and limit behavior, let us start with the linear range neutral steer arrangement from Problem 1. At the limit there is significant load transfer across the front axle. The front tires were equally loaded for the linear range calculation of Problem 1. Load transfer at the limit will result in reduced lateral force capability for the front axle due to tire load sensitivity. The rear tire's lateral force capability remains unchanged. This says that the vehicle will plow (push) in the nonlinear tire range and at the limit.

   In an actual vehicle the addition of an anti-roll bar will induce second-order effects based on the roll gradient. This is due primarily to camber change, assuming that the front suspension is an independent.

5. In a 1g stop the longitudinal reaction force at the CG is equal to the weight of the vehicle, in this case 2000 lb. By taking moments about the rear axle the longitudinal load transfer is determined as:

$$\Delta F_{zF}\ell = A_x W h \quad \Rightarrow \quad \Delta F_{zF}(100) = 2000(15) \quad \Rightarrow \quad \Delta F_{zF} = 300 \text{ lb.}$$

Thus, in a 1g stop the front wheel loads will increase by 300 lb. over the static wheel load (150 lb./wheel) and the rear wheel load will decrease by 300 lb. from the static condition. The CH10 with its CG located as in Problem 1 has all three wheels with identical static wheel loads: 667 lb.

Dynamic wheel loads for a 1g stop then are:

$$F_{zLF} = F_{zRF} = 667 + 150 = 817 \text{ lb.}$$

$$F_{zR} = 667 - 300 = 367 \text{ lb.}$$

Now assume the tractive friction coefficient is independent of load. This allows the brake distribution to be set in proportion with the load distribution. The front/rear load distribution is $2(817)/367 = 4.45/1$. The front brake proportion then becomes:

$$\% \text{ Front Brake} = \frac{4.45}{4.45 + 1.0} \times 100\% = 81.6\%$$

Any brake balance above 81.6% will lock the front wheels first in a 1g stop, provided that the front brake is split equally between the two front wheels. Note that this simple calculation is not the calculation required to determine the brake balance for *maximum* deceleration which requires data on the load sensitivity of the tire longitudinal coefficient of friction[1].

6. Refer to RCVD Figure 17.12 and the anti-dive equation on RCVD p.618 to design for 50% anti-dive:

$$\% \text{ anti-dive} = \frac{W\left(A_X/g\right)(\% \text{ front braking})\left(\text{svsa height}/\text{svsa length}\right)}{W\left(A_X/g\right)\left(h/\ell\right)}$$

$$\% \text{ anti-dive} = (\% \text{ front braking})\,(\tan\phi_F)\left(\ell/h\right)$$

$$50\% = (80\%)\,(\tan\phi_F)\left(100/15\right) \quad \Rightarrow \quad \phi_F = 5.35 \text{ deg.}$$

The instant center of the side view swing arm must be located on a line inclined 5.35 deg. with its origin at the front tire contact patch as shown in Figure 10.3.

[1] Maurice Olley used an interesting, graphical approach to determine brake distribution. See *Chassis Design: Principles and Analysis* by William F. and Douglas L. Milliken, SAE, 2002, Section 9.3.

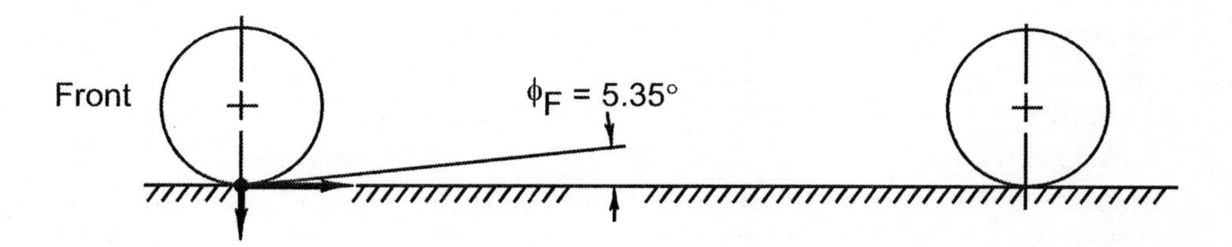

Figure 10.3 *Sketch showing swing arm for 50% anti-dive.*

7. In the absence of aerodynamic downforce the relationship between the static wheel loads and the cornering stiffnesses for neutral steer is given by:

$$\frac{F_{zR}}{W} = \frac{a}{\ell} \quad \Rightarrow \quad \frac{F_{zR}}{W} = \frac{C_R}{C_F + C_R}$$

Be careful when applying this equation in the presence of aerodynamic downforce. First, note that the ratio of front wheel load to total weight is taken at static conditions and is equal to the ratio of lengths a and ℓ. This value *does not change* with aerodynamic downforce and is therefore speed independent. The cornering stiffnesses, however, are a function of the dynamic wheel loads, including any additional loads caused by aerodynamics.

The wheel loads at this (unspecified) speed are:

$$F_{zLF} = F_{zRF} = \frac{1}{2}\left(W\frac{b}{\ell} + 500\right)$$

$$F_{zR} = W\frac{a}{\ell} + 500$$

Cornering stiffnesses at the front and the rear are evaluated at these loads. Thus, the formula for neutral steer behavior can be written as:

$$\left.\frac{F_{zR}}{W}\right|_{\text{static}} = \frac{a}{\ell} = \frac{C_R|_{((Wa/\ell)+500)}}{2C_F|_{(((Wb/\ell)+500)/2)} + C_R|_{((Wa/\ell)+500)}}$$

This is the same solution task as in Problem 1. In Problem 1 the solution was obtained through the use of a table. Here, a polynomial will be fit to the data to arrive at the solution algebraically. The cornering stiffness curve in Figure 10.2 can be represented by the following function, derived via a least-squares fit:

$$C(F_z) = \left(-5.69 \times 10^{-5}\right) F_z^2 + 0.385F_z - 1.355$$

This cornering stiffness expression can be substituted into the previous expression for neutral steer and solved for the CG location. The result is $a/\ell = 0.431$. The CG should be located 43 in. behind the front axle, 10 in. farther rearward than in the static calculation of Problem 1. This is a very big change, indicating an aero imbalance. The downforce distribution should be shifted forward if the CG is to be left at its static (33 in.) location.

This question hints at the importance of designing the correct progression (against speed) of front/rear aerodynamic distribution into the vehicle shape. The speed dependence of the aerodynamic distribution should be complementary to the non-linearity of the tire cornering stiffness curves to maintain neutral steer at all speeds (if possible).

8. The four-wheeled vehicle experiences load transfer across both axles in a proportion determined by the roll stiffness distribution. Since the rear wheel on the CH10 could not provide any roll stiffness, the calculations were significantly simplified when it came to oversteer and understeer considerations.

CHAPTER 11

Testing and Development

Solutions to this chapter's problems begin on page 144.

11.1 Problems

1. You are the designer of a new open wheel race car for a professional racing series. Now that the car is complete, what items will you measure in the shop to confirm your chassis and suspension design calculations? Because of a tight schedule, assume that all the measurements must be made in three working days.

2. Prior to the initial track test of the above car, what data will be required by the race engineer?

3. In his book *Drive to Win*[1], Carroll Smith gives a very good description of what top racing drivers do to get the best performance from their team members. Produce a short outline that recasts this in terms of race engineer behavior at a test session.

[1]*Drive to Win*, Carroll Smith Consulting, Inc., ISBN 0-9651600-0-9. 1996.

11.2 Simple Measurements and Experiments

NOTE: All experiments should be carried out at low speeds (say, below 30 mph) under safe conditions. A shopping market or campus parking lot early on Sunday mornings often makes a good skidpad.

WARNING: Skidpad testing can produce significant amounts of tire wear in the shoulder area of the tires. Do not continue the experiments for many laps unless you are unconcerned about tire wear. Tire wear can be reduced by performing the experiments on wet or snow-covered surfaces.

1. Refer to Questions 1 and 2 (and their answers) for a list of measurements to make on your race car. Some of these measurements will be difficult or impossible to make without access to special equipment, but others can be measured quite easily with a little ingenuity. Make as many measurements as you are able.

2. Locate a parking lot or other large, smooth, paved area. Mark a measured circle with chalk or cones and perform constant radius skidpad tests on your race (or street) car as described in RCVD Section 11.7, p.383-5. Add a protractor to the steering wheel and attach a simple pointer (made from soft, safe materials) to the car to measure steering wheel angle. Use a stopwatch to time laps and then calculate the lateral acceleration. Plot the data as in RCVD Figure 11.2, p.384.

11.3 Problem Answers

1. Some relatively easy-to-make chassis measurements include:

   - Weight and balance information including CG longitudinal and lateral location. If possible, CG height should also be determined.
   - Chassis torsional stiffness. To measure this, apply the input torque at the wheel center locations, with springs/dampers replaced by solid members.
   - Ride steer (bump steer) and ride camber (both measured with springs removed).
   - Actual steering ratio for both wheels (steer-steer test on Weaver/alignment plates).
   - Actual installed spring rates and anti-roll bar rates—these are likely to be non-linear. If possible, test with more than one spring rate and anti-roll bar setting (or rate).

- Calibrate vehicle ride height with bump-stop location and bump-stop spring rate. This is important since bump "rubbers" are often used to control ride height at high speed under aerodynamic download, which tends to be ride height sensitive.

If you have access to a kinematics and compliance rig, many more tests can be made in a short time. Kinematics (suspension geometry) measured with the actual spring loads present will never agree exactly with the predictions from drafting layout or simple geometry software. For race cars, compliances (measured with force and torque inputs at the tire contact patches) should be minimal.

2. In addition to the items mentioned in the answer to Question 1:

    - Individual static wheel loads (wheel scale loads) measured on a known flat surface.
    - Suggested wheel alignment: caster, camber, toe.
    - Aerodynamic data as a function of wing angles, as well as body ride height and pitch angle. These are typically available from wind tunnel tests.
    - Any tire data that are available. In many cases this will be limited to tire spring rate. Force and moment data versus load and slip angle is still quite hard to get.
    - Damper curves and other manufacturer's data.
    - Engine power curves and gear ratio charts are essential. Some data on the braking system may be helpful if there is a problem with brake balance.

    In general, it is rare to have "too much" data. Usually the problem is just the opposite and decisions must be made quickly without sufficient data.

3. Some thoughts include:

    - Keep calm. Presumably this will be easier to do if you are familiar with the driver, crew and all the various systems on the car.
    - Have a plan going in, but be prepared to deviate from it for good reasons.
    - Recognize that there are a number of different specialists who are all vying for some time to test their systems—engine, transmission & cooling, chassis, aerodynamics, etc. Communication is necessary to make sure that each specialty gets their required test data.
    - It is important that the car be driven consistently near the limit. Make this clear to the driver.
    - Question the driver in detail during and after the test session to get as much feedback as possible on the actual behavior of the car, particularly at any problem spots on the track.

11.4 Comments on Simple Measurements and Experiments

1. Formula SAE (and other collegiate competition) teams are especially encouraged to measure as many aspects of their car as possible. While design calculations are important, it is also important to be able to prove to the judges that your vehicle meets your design goals.

2. Low speed, constant radius tests are relatively easy to run and give good objective data on car behavior under controlled conditions. Even a very simple test as described here will give useful data on:

   - Ackermann steering wheel angle, steering ratio and left-right symmetry
   - Linear range stability, understeer/oversteer
   - Limit behavior (plow/spin/drift)
   - Longitudinal trim change behavior (dropped throttle and braking)

   Circular skidpad testing is typically run at lower speeds which presents some problems when testing faster cars. For example, the aerodynamic forces (drag and downforce) will not be correct. In this case (faster car) actual race track testing will be required.

CHAPTER **12**

Chassis Set-up

Solutions to this chapter's problems begin on page 149.

12.1 Problems

1. Assume that you are the race engineer for a rear drive car with large, lightly loaded tires. In a low speed, unbanked corner the driver complains of steady-state push. Changes to the anti-roll bars both front and rear have not significantly altered the situation.

   a. Explain what may be happening.

   b. Propose one or more other avenues to explore.

2. Consider the F1 vehicle denoted as Second Example on RCVD p.257. The calculation given there shows that 3 deg. at the steering wheel, δ_{SW}, produces a 1g turn at 180 mph. Suppose that the driver reports that the car is "too sensitive" to steering inputs. Consider the following three modifications:

   - Reducing the steering gear ratio by 10%, from 12:1 to 13.2:1.

- Reducing both the front and rear tire cornering stiffnesses by 10% so that $C_F = -977.4$ lb./deg. and $C_R = -1517$ lb./deg.
- Increasing the wheelbase by 10% from 9.52 ft. to 10.47 ft. while keeping front/rear weight distribution at 40/60.

Using a two-degree-of-freedom bicycle model, calculate the vehicle yaw rate gain r/δ_{SW} for the baseline case of the Second Example car, as well as for each of the three proposed modified versions of it. Which modification reduces the steering gain the most? Discuss and justify your answer.

3. As a race engineer for an open-wheel road race car, your driver requests one of the following changes in the car set-up. Qualitatively discuss the primary effect(s) of each change, consider what secondary or side effects may occur and note any limiting (or boundary) conditions that may be involved:

   - Reduce front wing angle (make the wing chord more horizontal).
   - Increase the negative camber on the front wheels.
   - Stiffen the rear springs.

   Now consider the same changes on an oval track sedan (the front wing is now a front air dam).

12.2 Simple Measurements and Experiments

> NOTE: All experiments should be carried out at low speeds (say, below 30 mph) under safe conditions. A shopping market or campus parking lot on early Sunday morning often makes a good skidpad.
>
> WARNING: Skidpad testing can produce significant amounts of tire wear in the shoulder area of the tires. Do not continue the experiment for many laps unless you are unconcerned about tire wear. Tire wear can be reduced by performing the experiments on wet or snow-covered surfaces.

1. Refer to Experiment 2 in Chapter 11 and RCVD Chapter 12. Make a series of changes in the set-up of your car. Document each change and perform a constant radius test to determine the effect of the change.

12.3 Problem Answers

1. a. Anti-roll bars typically affect the total axle lateral force through tire load sensitivity—which is seen as nonlinearity in a plot of tire lateral force against load (at a range of constant slip angles). With the "classic" use of an anti-roll bar, one end of the car resists more of the lateral load transfer than the other. The resulting differences between the tire loads effectively reduce the lateral force available (when compared to the more-equally loaded tires on the other end of the car). There are at least three things that could be occurring with a symmetric car (asymmetric cars for oval tracks raise another set of issues):

    - If the tires are lightly loaded, the amount of tire load sensitivity may be so slight that taking all of the lateral load transfer on one end of the car may not degrade the lateral force performance significantly.

    - If the roll centers are located well above the ground (perhaps with an axle suspension), the rolling moment about the roll axis will be small. Thus the roll angles will be small and anti-roll bars will not have a large effect on tire loads. In this case the load transfer distribution is controlled by the relative heights of the roll centers, not by the relative roll stiffnesses of the front and rear suspensions.

    - If the car is running a locked differential (spool) or is fitted with a limited slip differential with preload, the resulting moment from the rear axle will tend to bias a car toward push.

   b. The adjustment(s) needed to balance the car depend on the cause of the steady-state push.

    If the cause is a lack of tire load sensitivity with lightly loaded tires, some changes that may have an effect include: adjusting tire pressures, changing tire size/width (if possible put wider or higher "grip" tires on the front) and moving the CG location rearward.

    If high roll centers are making the anti-roll bars ineffective, adjusting the roll center heights (lowering the front or raising the rear) may have an effect.

    If the stabilizing torque from a locked differential is the root cause, consider freeing up the differential if wheelspin on corner exit is not a large problem. This can be done by either changing from a spool to a limited slip differential or by reducing the clutch-pack-preload if a limited slip differential is already in use.

2. As a measure of "sensitivity" choose the yaw rate sensitivity (gain) r/δ. The following expression can be derived from RCVD Equation 5.15, p.155, by substituting values for the stability derivatives given in RCVD Equation 5.9:

$$\left.\frac{r}{\delta}\right|_{ss} = \frac{\ell V C_F C_R}{(aC_F - bC_R)\, mV^2 + \ell^2 C_F C_R} \text{ (rad./sec.)/rad.}$$

This is the yaw rate sensitivity at the front wheels in (rad./sec.)/rad. To arrive at the yaw rate sensitivity at the steering wheel in (rad./sec.)/deg., divide by the steering ratio, G, and 57.3 (rad./deg.):

$$\left.\frac{r}{\delta_{SW}}\right|_{ss} = \frac{1}{57.3G}\frac{\ell V C_F C_R}{(aC_F - bC_R)\, mV^2 + \ell^2 C_F C_R} \text{ (rad./sec.)/(deg. SW)}$$

The baseline case from the Second Example with a 12:1 steering ratio gives:

$$\left.\frac{r}{\delta_{SW}}\right|_{ss} = \frac{25.563}{57.3 \times 12} = 0.03718 \text{ (rad./sec.)/(deg. SW)}$$

a. Reducing the steering gear ratio by 10% to 13.2:1 should give a 10% reduction in yaw rate sensitivity since yaw rate sensitivity at the steering wheel is proportional to steering ratio. With a 13.2:1 steering ratio:

$$\left.\frac{r}{\delta_{SW}}\right|_{ss} = \frac{25.563}{57.3 \times 13.2} = 0.03380 \text{ (rad./sec.)/(deg. SW)}$$

The reduction in yaw rate sensitivity is:

$$\frac{0.03718 - 0.03380}{0.03718} \times 100\% = 9.1\%$$

b. Reducing both C_F and C_R by 10% gives:

$$\left.\frac{r}{\delta_{SW}}\right|_{ss} = \frac{25.337}{57.3 \times 12} = 0.03685 \text{ (rad./sec.)/(deg. SW)}$$

Compared with the baseline case this is a reduction of:

$$\frac{0.03718 - 0.03685}{0.03718} \times 100\% = 0.9\%$$

c. Increasing the wheelbase by 10% gives:

$$\left.\frac{r}{\delta_{SW}}\right|_{ss} = \frac{23.517}{57.3 \times 12} = 0.03420 \text{ (rad./sec.)/(deg. SW)}$$

This is a reduction of:

$$\frac{0.03718 - 0.03420}{0.03718} \times 100\% = 8.0\%$$

As expected, the steering change gives a 1:1 reduction in r/δ at steady-state because the system is linear and superposition holds. Decreasing the tire cornering stiffnesses by 10% produces a negligible change, provided they are both reduced *simultaneously*. If one is reduced more than the other the effect will be more pronounced because the US/OS characteristics of the vehicle will be changed. Reducing or increasing the wheelbase produces an almost linear change in r/δ for this particular vehicle.

Now consider the relative difficulty in making these changes to an actual car. Small changes in the steering ratio can be made by either changing the steering rack (or box) for another ratio or changing the length of the steering arm at the wheel hub. Both are relatively straightforward mechanical changes. On the other hand, changing the wheelbase by 10% (approximately 14 in.) will most likely require a major redesign of the whole car including the aerodynamic package. The practical choice (of the choices used for this simplified problem) is to change the steering ratio.

Finally, while this solution uses r/δ (yaw rate gain, linear with speed) as a metric, there is no guarantee that this will correlate with the driver comment of "too sensitive". Other possible metrics include:

- Lateral acceleration gain, A_Y/δ, which is a function of u^2.
- Control gain, N/δ, which is a function of front cornering stiffness and the distance from the front wheels to the CG.

3. The following are some thoughts, but a number of different answers are correct, depending on the initial conditions:

   a. In general, reducing front wing angle will reduce front downforce, tending to change the balance toward "push". At the same time, it should reduce drag, allowing a higher top speed. If the wing was initially at a very high angle of attack (stalled), then reducing the angle of attack may reduce the drag significantly.

Secondary effects will be higher front ride height with possible changes in alignment (camber, toe, caster). Raising the front of the car changes the "rake angle" which can affect the overall aerodynamic performance, most notably by changing the contribution from the bottom of the car (which operates in ground effect).

b. Negative camber will increase front lateral force capability (up to a point) tending toward "spin" (or reduced "plow"). Secondary effects may include more tire wear (higher tire temperature) on the inner shoulder of the tire and decreased braking capability. Tire aligning torque characteristics change with camber and the steering geometry (toe, caster) may need to be adjusted to compensate.

c. Stiffer rear springs will increase the fraction of the lateral load transfer carried by the rear track. In general, this will reduce the cornering capability of the rear tires, tending toward "spin". Note, however, that with the very large tires in use on the rear of some cars, this "primary" effect may actually be very small. Stiffer springs also change the rear ride height at any given speed which can result in a change in the front/rear aerodynamic distribution.

If the car is a large sedan the coupling effects between ride height and aerodynamic downforce are less significant, but still important (since these cars are highly optimized for each track). Such a car may be more sensitive to chassis changes that affect tire loads since the tires are more heavily loaded—tire load sensitivity is more apparent.

12.4 Comments on Simple Measurements and Experiments

1. The set-up guides in RCVD Chapter 12 are meant only as general guidelines. If your car does not respond as predicted by the set-up tables, we recommend that you continue testing until the mechanism(s) at work have been identified and quantified.

CHAPTER **13**

Historical Note on Vehicle Dynamics Development

Solutions to this chapter's problems begin on page 155.

13.1 Problems

1. The Bugatti Type 35 racing car was first introduced in 1924 at the French Grand Prix at Lyon. It was the predecessor of the highly developed Type 51 of 1931 and set the stage for the Type 59, 1934. The Type 35 was campaigned by amateur as well as professional drivers and is reputed to have won more races than any other type of racing car. It has a reputation for good handling which must derive from its basic design, since extensive chassis tuning was not prevalent during its racing era.

   Some of its design characteristics (see RCVD Figure 13.1, p.414) are listed below:

   - All up racing weight $\sim$1850 lb., 40/60 front/rear weight distribution
   - Equal sized tires all around, 28 in. diameter, about 30 psi front, 35 psi rear
   - Solid axles front and rear with roll centers at about axle height

- Stiff ride springs, considerably stiffer on the front than on the rear, damping by friction shocks and inter-leaf friction
- Fast steering ratio, worm and wheel steering box, drag link to right front wheel and track rod, 6-8 deg. positive caster
- Positive static camber (wheels tilted outward) at front
- Front engine bolted solidly to frame and very substantial dash bulkhead, a very stiff chassis (torsional and beaming) for its day

Given the above, discuss in qualitative terms your estimate of the behavior of the Type 35 in the following situations:

a. Acceleration
b. Braking
c. Turn-in and steady cornering
d. Directional response to control
e. Rough road

Under what circumstances would you expect wheel lift on the front, trailing throttle oversteer, or breakaway on the rear?

Bill Milliken relates his experiences with the Type 35 in the answer to this question.

2. It is proposed to design a variable stability automobile (VSA) to simulate the steady-state characteristics of other cars in the linear range of tire performance by steering the front and rear wheels with a position servo control. The steering of the front wheels will be the result of signals from the driver plus other signals from sensed motion of the car and/or its components. The steering of the rear wheels will be based on signals from the sensed motion of the car and/or its components only.

   Using the concept of stability derivatives as discussed in RCVD Chapter 5, pp.152-168, and the discussion of variable stability in RCVD Chapter 13, pp.437-439, answer the following:

   a. Is it possible to change the understeer/oversteer of the VSA as measured at the steering wheel by steering only the front wheels (over and beyond driver input)? If so, what force or moment on the vehicle is affected? What primary motion variable should be sensed? What derivative is involved?

   b. Answer Question 2a for the case where understeer/oversteer is measured at the front wheels.

c. Is it possible to change the effective yaw damping of the VSA by steering only the front wheels (over and beyond driver input)?

d. If so, what force or moment on the vehicle is affected? What primary motion variable should be sensed? What derivative is involved?

e. The lateral damping derivative is Y_v (or Y_β for constant forward speed). How would one change the effective lateral damping using the control system on the VSA without introducing the effect of other derivatives?

f. When the under/oversteer, yaw damping and lateral damping effects are changed by steering of the front and rear wheels, would you expect the roll motion to be affected?

g. What servo would you introduce to gain an independent control of the rolling motion? What primary variable would you sense to control roll position and roll damping?

h. Suppose you wished to simulate a change in yaw inertia. What force or moment would you control? What primary variable would you sense? What derivative is involved?

i. Discuss the simulation of changes in lateral force and aligning torque compliances (see RCVD Chapter 23). What forces or moments are involved? What variables should be sensed? What derivatives are involved?

j. In addition to simulating the characteristics of other cars with the VSA, would you expect to be able to create behavior that cannot be attained by conventional, fixed-configuration automobiles?

13.2 Problem Answers

1. In answering this question, Bill Milliken first gives his impressions of his Type 35A which he used for personal transportation as well as racing. These are followed by an explanation of the possible basis for the impressions.

   a. Acceleration: Wheelspin was seldom encountered on a dry surface, but occasionally occurred on gravel or low-coefficient surfaces. Directional control was good with rapid response, so loss of control (spin) was seldom a problem during straight ahead acceleration.

Considering the power-to-weight ratio and load on the rear wheels (static plus the increase due to longitudinal acceleration), one would not expect rear wheel-spin on high-coefficient surfaces.

b. Braking: By modern standards, braking was just adequate. Directional control during braking was satisfactory, but pedal forces were high. With better brakes I might not have lost it at "Milliken's Corner", Watkins Glen Grand Prix, 1948!

The brakes were unboosted, cable operated with a mechanical equalizer in the cockpit. The brakes were small internal drum brakes with two shoes, non-self-energizing. Because of the pedal arrangement, left foot braking was occasionally used. Good stability under braking is accounted for by the dynamic wheel loading, and the near neutral response of the car due to higher rear tire pressure, larger lateral weight transfer on front and the positive camber of the front wheels all compensating for the basically aft CG.

c. Turn-In: Under hard braking, turn-in frequently resulted in breakaway at the rear wheels, which could usually be caught by reverse steering. Movies of the car at Watkins Glen (1948), Bridgehampton (1949), etc., consistently indicated this behavior on dry roads.

This occurred mainly because of the unloading of the rear wheels. On a tight circuit like Palm Beach Shores this turn-in behavior was desirable.

d. Steady Cornering: At high lateral accelerations, the steering control forces were very high. On Big Bend at Watkins Glen at 90-100 mph, I was tugging with both hands on one side of the wheel to keep from running off the outside of the road.

The high steering forces are due to large caster, the very fast steering ratio and friction in the system.

e. Directional Response to Control: The impression in driving was one of good response to the steering and well-damped transients.

Part of this feel may have been due to the fast steering. For its day, the car was not undertired. It would be very interesting to measure some transient responses to step steering inputs.

f. Rough Road: The car gives the impression of being very stiffly sprung, particularly on the front. At Giant's Despair (a hill climb), the photos show the front end off the ground on a rough turn which confirms driver impression of loss of control. On rough surfaces the front end is bobbing up and down continuously. Subjectively this gives a nice feeling of "liveliness" and the necessity for continuous control action, bringing the driver into the control loop. Objectively it is probably less desirable.

The answer to the question of wheel lift, trailing throttle oversteer and breakaway at the rear is a high lateral acceleration turn-in, particularly on a rough surface.

2. a. Understeer gradient is measured at the steering wheel as lateral acceleration per steering wheel angle in units of deg./g. A control algorithm can be written to steer the front wheels an amount more or less than that with a fixed steering ratio.

   For example, reducing the amount the front wheels steer at a given steering wheel angle results in a smaller lateral force from the front tires. This leads to a smaller lateral force from the rear tires at steady-state. The result is a smaller lateral acceleration at the given steering wheel angle (due to the smaller front wheel steer angles) and an increase in understeer gradient.

   This algorithm could be a function of steering wheel angle, lateral acceleration, yaw rate and/or others.

   The derivative N_β, static directional stability, is usually associated directly with over/understeer. The control algorithm does not change N_β (as defined for the Bicycle Model), but the understeer as experienced by the driver changes. By modifying the relationship between the driver control and the front and rear wheel steer angles a VSA can modify various handling characteristics including understeer. The inherent characteristics of the passive vehicle, however, remain unchanged.

   b. If understeer gradient is measured at the front wheels the understeer cannot be modified by steering the front wheels alone.

   c. Yes. Since yaw damping is a force proportional to yaw velocity a steering gain could be applied based on measured yaw rate.

   d. The derivative involved is the yaw damping derivative N_r. Yaw rate would be sensed and the front wheels steered appropriately to modify the yaw moment on the vehicle.

   e. Damping in sideslip can be altered by steering both front and rear wheels together in the proper amounts to give a lateral force proportional to lateral velocity. Either v or β would have to be sensed. Note that it is not always possible to decouple all the derivatives, so control affecting the damping in sideslip is likely to affect other vehicle characteristics.

f. Any of the previously considered control schemes will affect the roll motion of the vehicle.

g. Roll motion can be controlled by the use of servos between the sprung and unsprung masses such as at the shock absorber location, or a torque servo on the anti-roll bars. The primary variables to sense are position and velocity at the shock absorber. Vehicle lateral acceleration could also be sensed for more effective control.

h. To change the effective yaw inertia one must apply a yaw moment proportional to angular yaw acceleration, since $M_Z = I_Z\alpha$, where M_Z = yaw moment, α = angular acceleration and I_Z = yaw inertia.

For example, to simulate a decrease in yaw inertia the yaw moment would need to be proportionally increased through appropriate steering of the front and rear wheels.

i. Lateral force compliance implies a steer or inclination change due to a lateral force. Aligning torque compliance implies a steer or inclination change due to an aligning torque. Thus, suspension forces and moments should be sensed and appropriate steer commands generated to simulate the compliance steer or the effect of compliance inclination. If necessary, lateral acceleration can be measured instead of suspension forces to arrive at an approximate solution.

j. The VSA would be able to create behavior which cannot be achieved in passive vehicles. For example, a control algorithm could be written to hold steering sensitivity constant over a very large range of speeds and lateral accelerations.

CHAPTER **14**

Tire Data Treatment

Tabular tire force and moment data (typically recorded on testing machines) is voluminous and awkward to use in computer programs. One system for avoiding this problem is to nondimensionalize the data in a manner somewhat analogous to airfoil coefficients. The scheme referred to below and in RCVD Chapter 14 was initiated by Dr. Hugo Radt and is based on physically based tire theory. It enables a compression of the data to single curves for all loads tested. These in turn may be fitted by the "Magic Formula" or other types of curve fitting. From this the coefficients can be expanded for convenient use in vehicle models.

Solutions to this chapter's problems begin on page 161.

14.1 Problems

1. The table below lists values taken off a nondimensional tire data curve. What lateral force will this tire produce at a slip angle of 4 deg. and a normal load of 1000 lb.? At this load the friction coefficient is 0.965 and the cornering stiffness is −240 lb./deg. At what value of slip angle does the peak occur? What is the lateral force at the peak?

$\overline{\alpha}$	0.0	0.5	1.0	1.5	2.0	2.5	3.0	3.5	4.0
$\overline{F}$	0.000	0.462	0.763	0.912	0.974	0.996	1.000	0.995	0.988

2. Consider the lateral force and aligning torque data in the following tables:

Lateral Force (lb.)				
α	F_Z(lb.)			
(deg.)	700	1300	1900	2500
0	0	0	0	0
1	−478.6	−737.3	−855.4	−828.6
2	−864	−1340	−1581	−1572
3	−1113	−1742	−2102	−2171
4	−1248	−1967	−2424	−2604
5	−1308	−2074	−2598	−2889
6	−1329	−2115	−2680	−3061
7	−1331	−2122	−2709	−3156
8	−1323	−2113	−2711	−3202

Aligning Torque (lb.-ft.)				
α	F_Z(lb.)			
(deg.)	700	1300	1900	2500
0	0	0	0	0
1	32.14	92.73	158.7	204.9
2	49.44	145.4	261.3	362.2
3	47.98	145.1	279.8	435.9
4	37.68	116.9	241.9	428.6
5	26.83	85.24	187.6	374.3
6	17.97	58.55	137.2	305.2
7	11.19	37.81	95.96	238.9
8	6.043	21.91	63.31	181.3

a. Plot the given data.

b. Determine the cornering stiffness for each load.

c. Determine the friction coefficient at each load.

d. Determine the pneumatic trail for each load.

e. Use the results of 2b-2d to create tables of normalized lateral force and normalized aligning torque versus normalized slip angle.

f. Plot the results of 2e.

g. For each plot in 2f, determine the coefficients B, C, D and E which fit the Magic Formula to the data. Use the following guidelines:

- D is the peak value of the normalized curve.
- Assume C = 1.4 for the lateral force plot and C = 2.3 for the aligning torque plot.
- The product of B, C and D equals 1.0.
- E can be determined by trial and error.

h. Use the Magic Formula coefficients determined in 2g to expand the lateral force and aligning torque data at 1000 lb. normal load. Plot these curves with the data from 2a

14.2 Problem Answers

1. Start by calculating the normalized slip angle associated with a slip angle of 4 deg.:

$$\overline{\alpha} = \frac{C \tan \alpha}{\mu F_Z} = \frac{(-13751 \text{ lb./rad.}) \tan (4 \text{ deg.})}{(0.965)(-1000 \text{ lb.})} = 0.996$$

This value is approximately equal to 1.0, so the corresponding normalized lateral force can be read directly from the table. $\overline{F} = 0.7631$. Expanding this gives the lateral force at $\alpha = 4$ deg.:

$$F_y = \overline{F} \mu F_Z = (0.7631)(0.965)(-1000 \text{ lb.}) = -736.4 \text{ lb.}$$

The peak value of lateral force occurs at the peak value of normalized lateral force. This corresponds to a normalized slip angle of 3.0. Expanding gives the slip angle and peak lateral force:

$$\alpha = \tan^{-1}\left(\frac{\overline{\alpha}\mu F_Z}{C}\right) = \tan^{-1}\left(\frac{(3.0)(0.965)(-1000 \text{ lb.})}{-13751 \text{ lb./rad.}}\right)$$

$$\alpha = 0.2075 \text{ rad.} = 11.89 \text{ deg.}$$

$$F_y = \overline{F} \mu F_Z = (1.000)(0.965)(-1000 \text{ lb.}) = -965.0 \text{ lb.}$$

2. a. Plots of lateral force and aligning torque versus slip angle are given in Figures 14.1 and 14.2.

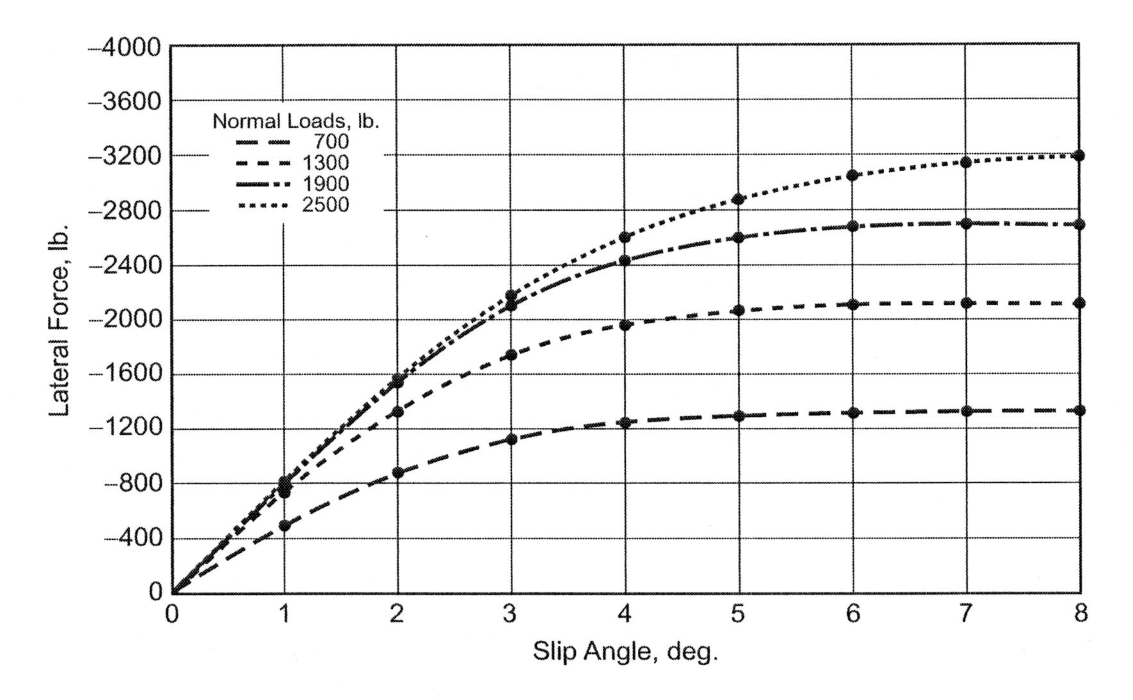

Figure 14.1 *Lateral force curves*

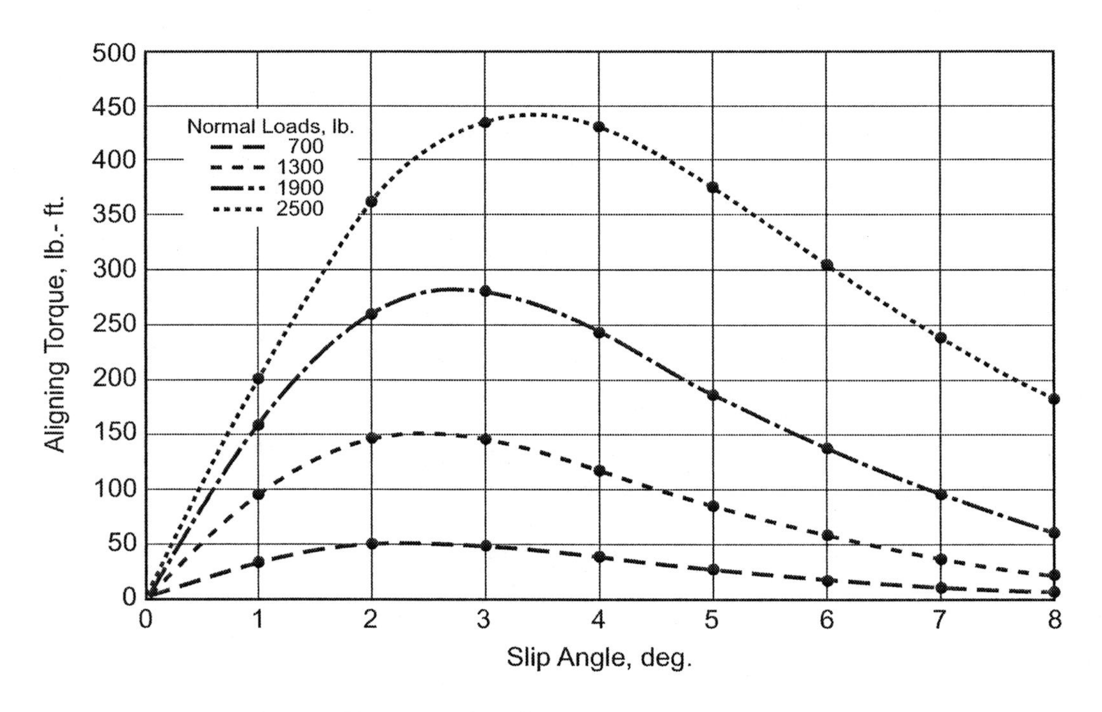

Figure 14.2 *Aligning torque curves*

b. Estimate cornering stiffness from the value at −1 deg. of slip angle:

F_Z (lb.)	700	1300	1900	2500
C (lb./deg.)	−478.6	−737.3	−855.4	−828.6
C (lb./rad.)	−27422	−42267	−49011	−47475

c. The friction coefficient is the peak lateral force divided by the normal load:

F_Z (lb.)	700	1300	1900	2500
μ (unitless)	1.901	1.632	1.427	1.281

d. Pneumatic trail can be estimated by dividing aligning torque by lateral force at −1 deg. of slip angle:

F_Z (lb.)	700	1300	1900	2500
T_Z (ft.)	−0.0672	−0.1258	−0.1855	−0.2473

e. Apply RCVD Formulas 14.1, 14.2 and 14.4 to the data given in the problem statement. For μ, C and T_Z use the values specific to each load.

Normalized Slip Angle, $\overline{\alpha}$				
α	F_Z(lb.)			
(deg.)	700	1300	1900	2500
0	0	0	0	0
−1	0.360	0.347	0.316	0.259
−2	0.719	0.695	0.631	0.518
−3	1.080	1.043	0.947	0.777
−4	1.441	1.392	1.264	1.037
−5	1.802	1.742	1.582	1.297
−6	2.165	2.092	1.900	1.558
−7	2.530	2.444	2.220	1.820
−8	2.895	2.798	2.541	2.084

Normalized Lateral Force, $\overline{F}$				
α	F_Z(lb.)			
(deg.)	700	1300	1900	2500
0	0	0	0	0
−1	0.360	0.347	0.316	0.259
−2	0.649	0.631	0.583	0.491
−3	0.836	0.821	0.775	0.678
−4	0.938	0.927	0.894	0.813
−5	0.983	0.977	0.958	0.902
−6	0.998	0.997	0.989	0.956
−7	1.000	1.000	0.999	0.986
−8	0.994	0.996	1.000	1.000

Normalized Aligning Torque, $\overline{M}$				
α	F_Z(lb.)			
(deg.)	700	1300	1900	2500
0	0	0	0	0
−1	0.360	0.347	0.316	0.259
−2	0.553	0.545	0.520	0.457
−3	0.537	0.554	0.556	0.551
−4	0.422	0.438	0.481	0.541
−5	0.300	0.319	0.373	0.473
−6	0.201	0.219	0.273	0.385
−7	0.125	0.142	0.191	0.302
−8	0.068	0.082	0.126	0.229

f. Plots of normalized lateral force and normalized aligning torque versus normalized slip angle are given in Figures 14.3 and 14.4.

g. To fit the normalized lateral force curve, the following coefficients work well: $B = 0.714$, $C = 1.4$, $D = 1.0$, $E = -0.45$.

To fit the normalized aligning torque curve, the following coefficients work well: $B = 0.8$, $C = 2.3$, $D = 0.55$, $E = -1.15$.

The Magic Formula line using these coefficients is drawn among the data in Figures 14.3 and 14.4.

h. To expand the data at $F_Z = 1000$ lb. normal load, the values of cornering stiffness, friction coefficient and pneumatic trail need to be estimated. Based on the data at the other loads, the following values can be assumed: $C = -645$ lb./deg., $\mu = 1.76$ lb./deg. and $T_Z = -0.097$ lb./deg. The result of the expansion is plotted in Figures 14.5 and 14.6.

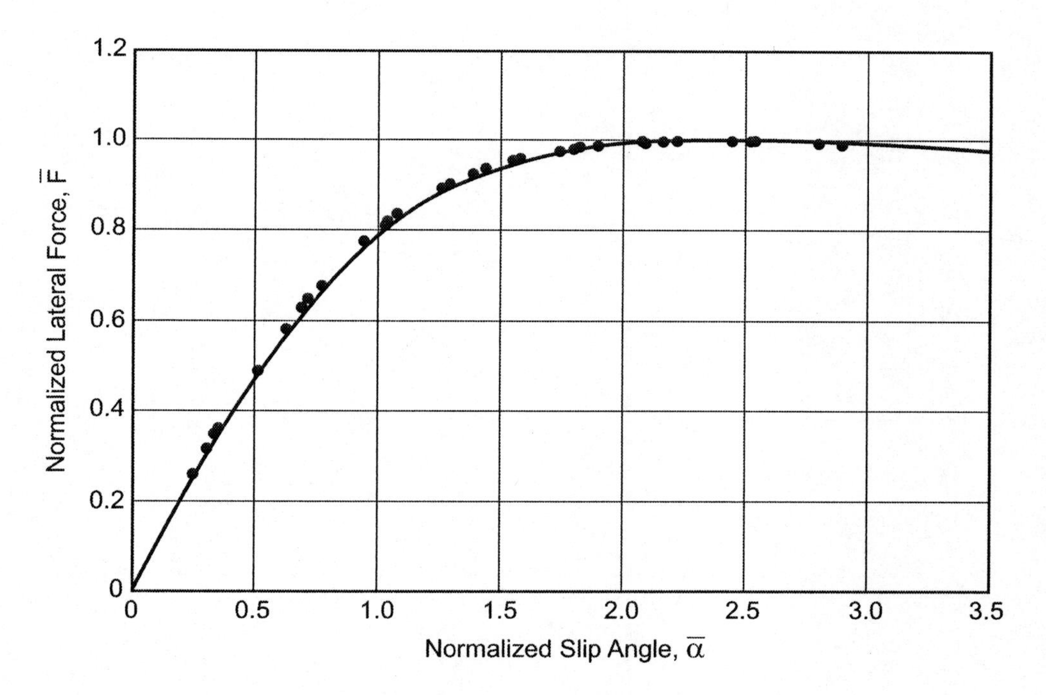

Figure 14.3 *Normalized lateral force*

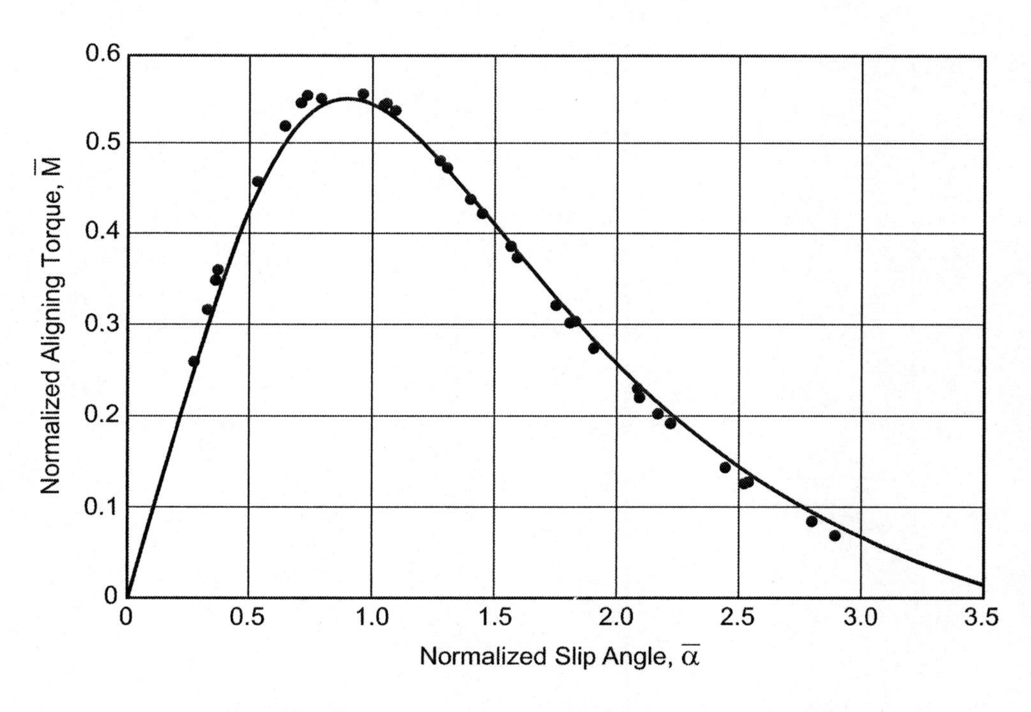

Figure 14.4 *Normalized aligning torque*

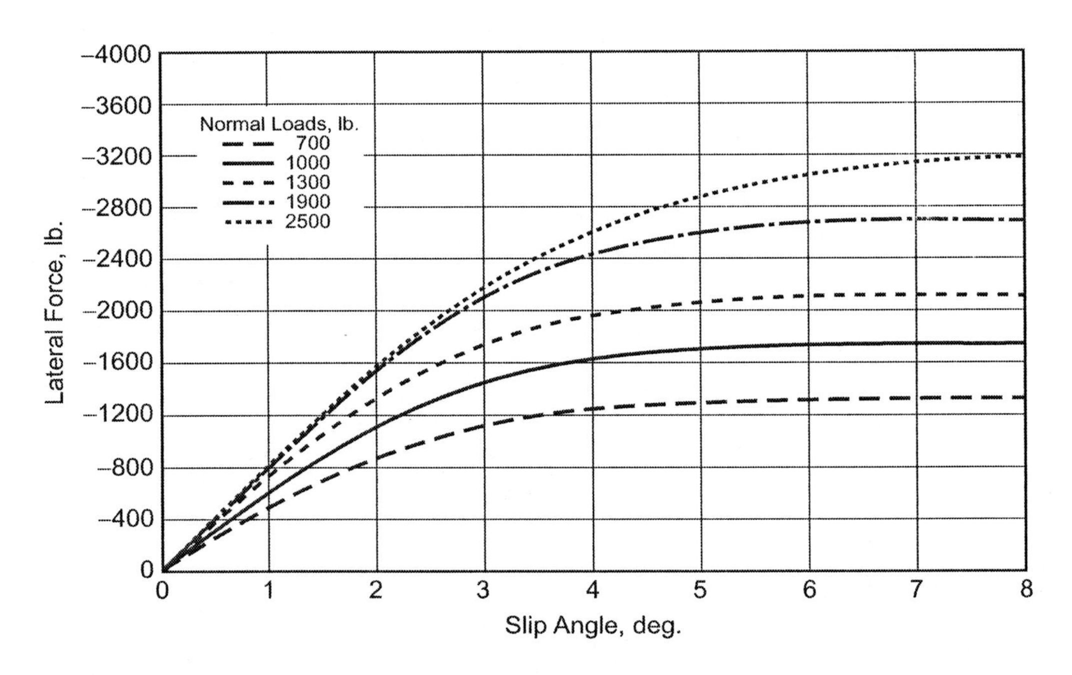

Figure 14.5 *Lateral force with expanded 1000 lb. load*

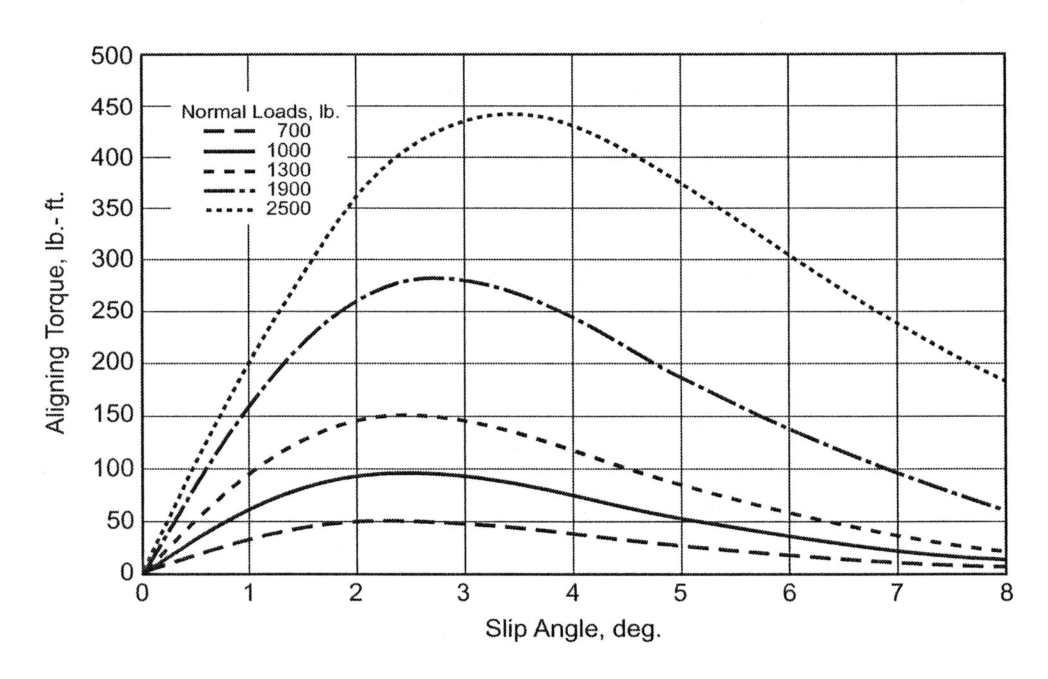

Figure 14.6 *Normalized aligning torque with expanded 1000 lb. load*

CHAPTER **15**

Applied Aerodynamics

Solutions to this chapter's problems begin on page 170.

15.1 Problems

1. Obtain an issue of *Racecar Engineering* or other motorsports publication which contains pictures of current Grand Prix cars. Identify the following aerodynamic flow control devices on the various cars, explain why they are used and briefly how they work:

   a. Guide vanes
   b. Fence
   c. Lip
   d. Gurney lip (wickerbill)
   e. Vortex generator
   f. Vent
   g. Scoop
   h. Diffuser
   i. Fillet
   j. End plate
   k. Slot/Slat
   l. Fins

2. Consider a flat plate placed perpendicular to the airstream as in RCVD Figure 15.68, p.563. The total drag force and drag coefficient of the plate depend not only on the stagnation pressure, but also on the edge conditions and the pressure on the back surface of the plate. Discuss whether or not a racing car could have a total drag, based on frontal area, greater than that of a flat plate with equal frontal area.

3. [1]In 1978 Brabham built the noteworthy "Fan Car" which went on to win the Swedish Grand Prix before being outlawed. Why did this car have such an advantage in cornering?

4. The exhaust and exhaust pipes play interesting aerodynamic roles in various classes of auto racing. Explain the reasoning for the exhaust location and direction on the following race cars:

   a. Top Fuel or Funny Car dragsters

   b. Formula One, CART or IRL chassis

   c. NASCAR sedans

5. Discuss the advantages and disadvantages of applying aerodynamic loads directly to the wheel hub or hub carrier. Compare against the generally mandated practice of applying aerodynamic loads to the chassis.

6. a. Calculate the lateral force produced by a pair of 12 $ft.^2$ sideboards as shown, for example, in RCVD Figure 15.66, p.562. Consider the case when a Supermodified with these sideboards is cornering at 130 mph with a vehicle sideslip angle of 7 deg. Assume the data in RCVD Figure 15.69 p.563 is valid for these sideboards.

   b. Based on your answer to 6a, calculate the roll moment from this lateral force. Assume the vehicle roll axis is on the ground and that the centroid of the sideboards is located 5 ft. above the ground. Note that this roll moment opposes the roll moment due to centrifugal force.

---

[1]More information on this car can be found in *Formula 1 Technology* by Peter Wright, SAE, 2001, p.216.

7. *Automobile* magazine gives the following information on the 1997 Corvette. The tires are Goodyear Eagle F1 GS Extended Mobility (i.e., run-flat), 245/45ZR-17 on the front and 275/40ZR-18 on the rear.

Top speed	172 mph
Engine	345 bhp at 5600 rpm
Weight	3218 lb. (curb) 3540 lb. (gross, estimated)
C_D	0.29 (based on frontal area)
Front cross-section	See Figure 15.1

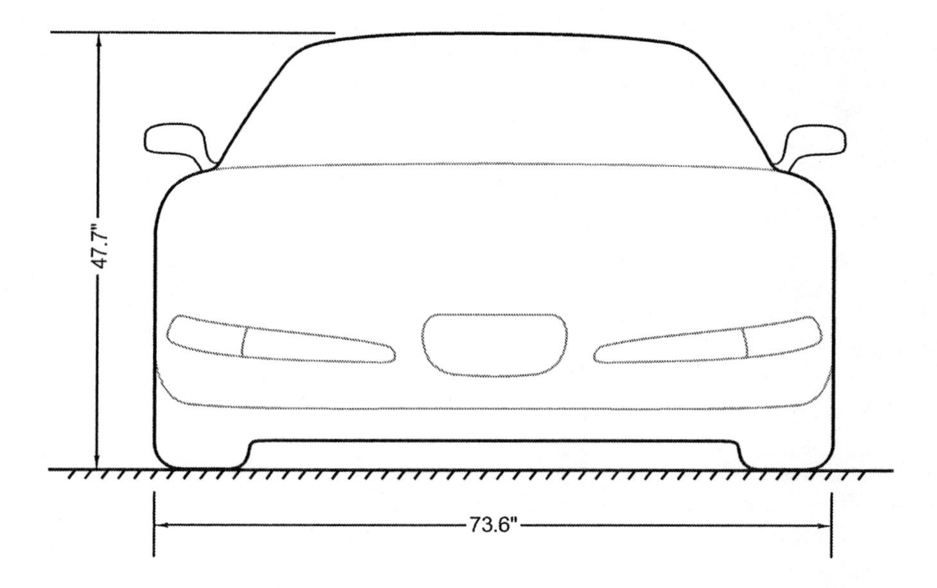

Figure 15.1 *Corvette cross-section*

Determine whether or not the top speed is consistent with the other data. Assume that the gearing is such that the engine power peak is reached at the quoted top speed (we have been assured by Chevrolet that this is true and that no speed-limiter is used).

Refer to pp.71-75 in RCVD Chapter 2 for insight on tire rolling resistance under driving torque. Assume that tire slip and camber are zero. No measured data on rolling resistance is available for these run-flat tires but it is known that, in general, the rolling resistance coefficient goes up rapidly with speed above 60-80 mph. The rise is slower for wide radial tires than for narrow bias tires.

15.2 Simple Measurements and Experiments

> NOTE: All experiments should be conducted legally and under safe conditions. While an expressway on Sunday morning (minimal traffic) or a lightly traveled back road work well, these tests may also be conducted on a constant radius circle as in previous chapters.

1. Aerodynamic testing normally involves wind tunnels and/or instrumented tests run on-road at high speeds. As a result, many racers automatically assume that aerodynamic testing is out of reach for cost reasons. However, simple tuft testing run at low speeds is available to any race team with time and a place to run the car. Tape simple yarn tufts on the surface of your car. Make the tufts about 3 in. long and place them on a 6 in. grid, closer in areas of sharp curvature or near duct work. Observe and videotape the tuft behavior from another vehicle driven near the test vehicle.

2. Review the data from tuft testing and locate any problem areas. Then modify the vehicle shape locally (or add aerodynamic devices) and re-test. There are a wide variety of aerodynamic devices available—these are listed in RCVD Chapter 15.

15.3 Problem Answers

1. In addition to the aerodynamic flow control devices labeled on Figure 15.2, see also RCVD Figure 15.54, p.548, which shows vents located on top of the wheel housing and RCVD Figure 15.57, p.551, which shows vortex generators.

2. There is nothing that prevents a moving body from having a drag coefficient greater than that of a flat plate with equal frontal area. Shaped bodies which have a nonzero characteristic length in the direction of V_∞ can produce lift (or downforce). Such a body does not really have "edges" in the same sense that the flat plate placed perpendicular to V_∞ does. Rather, there is a continuous flow distribution, pressure distribution and a boundary layer distributed along the characteristic length of the body.

   Referring to RCVD Figure 15.68, p.563, the frontal area of a race car has an h/b ratio of approximately 0.4. A flat plate with this ratio has a drag coefficient of about 1.18. Most Formula One, CART and IRL cars have drag coefficients greater than 1.0, approaching 2.0 in high downforce configuration.

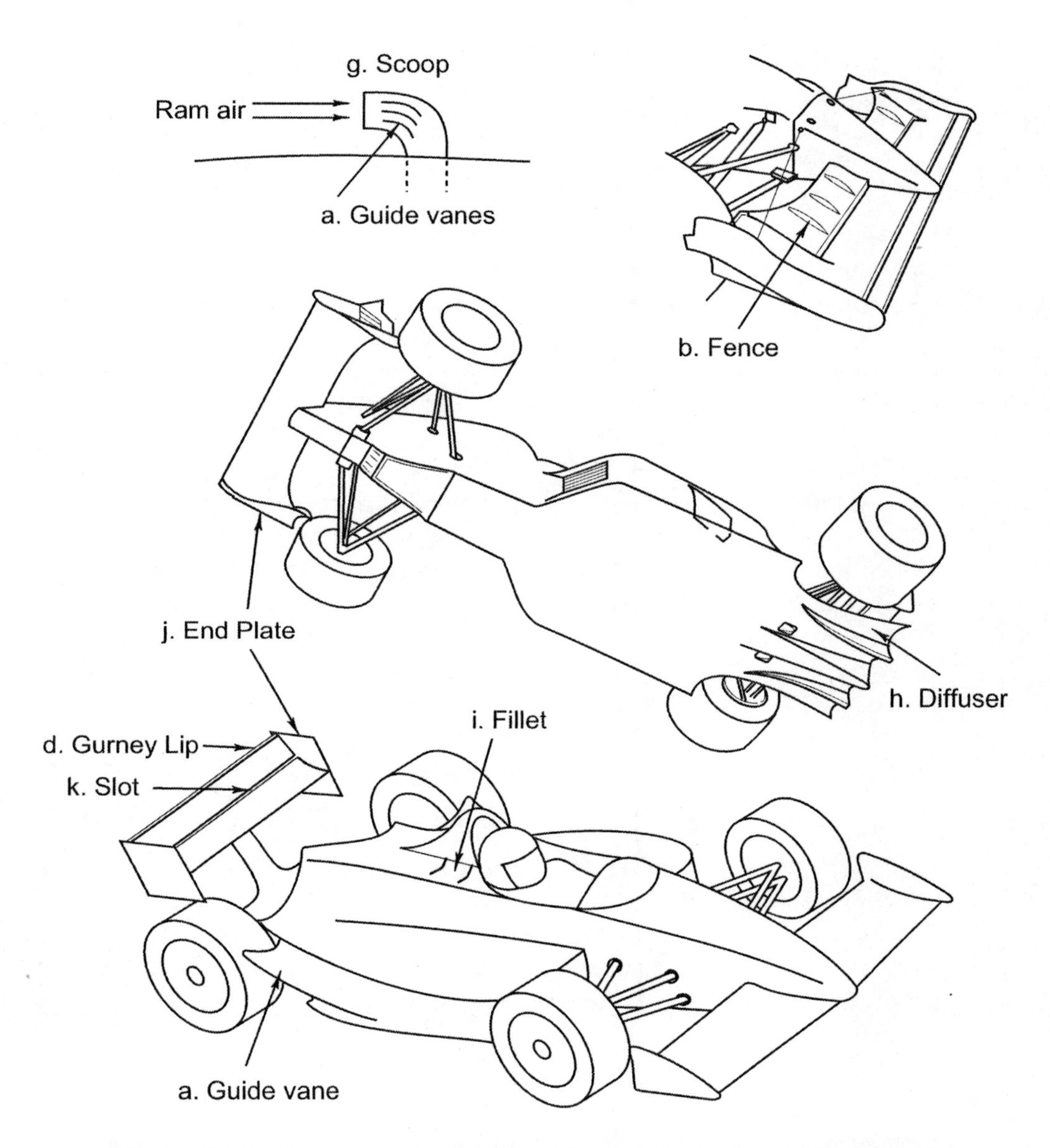

Figure 15.2 *Representative vehicles with aerodynamic devices labeled*

3. The Brabham used a large fan to extract air from underneath the car. This created an even lower pressure area underneath the vehicle, resulting in higher downforce. It was very much like the low pressure area created by a vacuum cleaner. Differential pressure creates force, so to get downforce either the pressure on the upper surface must increase or the pressure on the lower surface must decrease. Instead of creating low and high pressures with passive elements (wings, tunnels, overall body shape) the fan was used to create a low pressure area beneath the car.

As an aside, since fan speed was directly related to engine speed, drivers adapted to slowing before the corners and then taking the corners with as much throttle as possible. Lifting mid-corner would result in a decrease in engine rpm, fan rpm and thus downforce. When cornering at the limit the driver would have been ill-advised to lift off the throttle.

4. a. Dragsters typically have the exhaust just ahead of the rear wheels and pointing upward. The expanding exhaust gases provide downward thrust in the same way a rocket or jet engine produces thrust. The exhaust from a 6000 hp engine is capable of producing some amount of useful downward thrust, increasing tire loads and leading to increased tractive forces.

   b. F1, CART and IRL cars are highly dependent on aerodynamics. Exhaust location is partly determined by the need to minimize its effect on heat sensitive components. Beyond this, it is used to help modify the flow of air around the rear of the car. Whether exhausted below the car or through the upper surface, exhaust gases can be used to help extract air from underneath the car, improve tunnel efficiency or to make the rear wing more effective. Designs vary greatly, as a consensus has not been reached (not surprisingly considering the complexity of aerodynamics).

   c. NASCAR sedans typically exhaust out the left side of the car below the driver's door. A few teams prefer to exhaust out the right side, and they benefit from a very small amount of exhaust thrust helping in the (almost universal) left hand turns. The left side is preferred to keep the exhaust from bottoming on bumps while in a left hand turn (even though this places the exhaust directly beneath the driver's seat). While the exhaust does not play that much of a role aerodynamically, the presence of an exhaust pipe running half the width of the car serves to lower the pressure beneath the rear of the car by inhibiting flow past the exhaust pipe, possibly creating a little more downforce.

5. When wings first began appearing on race cars in the mid-1960s it was not uncommon to mount them high above the vehicle on "stilts" which connected directly to the suspension upright. By connecting to the upright or hub carrier, aerodynamic downforce could be used to increase wheel loads without influencing the sprung mass. Soft ride springs and "traditional" sprung mass frequencies and ride heights could be preserved. Unfortunately, the collapse of numerous suspension mounted wings caused several tragic accidents and led to an almost universal ban on suspension mounted aerodynamic devices.

   One of the most significant problems with suspension mounted wings is that the suspension needs to move. As such, a wing rigidly connected across the rear axle of a vehicle will be forced to flex with suspension travel. This can be overcome by placing air shocks in the mounting struts as done in Supermodifieds which still mount the main wing directly to the rear suspension[2]. Failure of the mounting struts is extremely rare.

Of course, it is much more difficult to pass the aerodynamic loads generated by the bodywork directly to the suspension. The Lotus T-88 Formula 1 car of 1981 managed to do this with a "dual chassis". It was subsequently outlawed[3].

6. a. RCVD Figure 15.69 indicates the lift coefficient $C_L = 0.09$ at an angle of attack of 7 deg. Since the plate is vertical it is interpreted as the side force coefficient. The 130 mph speed converts to 190.6 ft./sec. Calculation of the lateral force is as follows:

$$F_Y = 2\left(\frac{1}{2}\rho A C_L V^2\right) = (0.002378)(12)(0.45)(190.6)^2 = 466.8 \text{ lb.}$$

   The "2" in the lateral force expression accounts for the fact that there are a pair of sideboards, each with an area of 12 ft.^{2}. Note that this lateral force acts toward the center of the turn. On a 1600 lb. vehicle it contributes about 0.3g of lateral acceleration.

   b. With the roll axis on the ground, the moment from the lateral force produced by the sideboards is simply:

$$M_X = (5)(466.8) = 2334 \text{ lb.-ft.}$$

   There are two large rolling moments, one due to centrifugal force and one due to the sideboards. The sideboard moment reduces the adverse rolling moment of the centrifugal force. As such, it reduces the lateral load transfer in cornering, allowing for more equal wheel loads and resulting in increased cornering capability. Since the centroid of the sideboards is approximately three times as high above the roll axis as the CG, the moment from the sideboards cancels the moment from approximately 1400 lb. of centrifugal force (almost 1g's worth of lateral acceleration) at this speed. The roll moment provided by the sideboards is significant and leads to increased cornering speeds through the resulting reduction in lateral load transfer.

7. This problem is in the nature of engineering judgment where exact data is not available. Reasonable approximations will be required to reach a conclusion.

---

[2]See "Wacky Racers" by Edward Kasprzak, *Racecar Engineering*, Vol. 12, No. 12, December 2002.

[3]See "Grounded" by Peter Wright, *Racecar Engineering*, Vol. 11, No. 12, December 2001 or *Formula 1 Technology* by Peter Wright, SAE, 2001, p.217.

First, calculate the aero drag at the top speed of 172 mph. Figure 15.1 can be subdivided into a grid of 1 ft.^{2} squares. Counting squares gives an approximate frontal area of 19.8ft.^{2}. Assuming standard temperature and pressure:

$$\text{Air Drag, } D = C_D A q \text{ where } q = \frac{1}{2}\rho V^2 \approx \frac{V_{mph}^2}{391}$$

$$D = 0.29 \times 19.8 \times \frac{172^2}{391} = 434.5 \text{ lb.}$$

The power required to overcome this air drag is:

$$P_{aero} = \frac{DV_{mph}}{375} = \frac{434.5 \times 172}{375} = 199.3 \approx 200 \text{ hp}$$

Next, estimate transmission efficiency between engine and rear wheels at 90% (the combination of the overdrive top gear and the rear axle). The available power at rear wheels is reduced from that available at the engine by this percentage: $0.9 \times 345 = 310$ hp.

Of the 310 hp at the rear wheels, 200 hp goes into air drag. This leaves 110 hp which must go into the tires. Now perform calculations in terms of force instead of power. The thrust (tractive force) from the driving wheels at 172 mph is:

$$P = \frac{F_x V}{375} \quad \Rightarrow \quad F_x = \frac{375P}{V} = \frac{375 \times 310}{172} = 676 \text{ lb.}$$

The utilization of this force is expressed by the following equation, based on the discussion that starts on RCVD p.71. Since the car is at a steady top speed (not accelerating), the net longitudinal force on the vehicle is zero:

$$\sum(\text{longitudinal forces}) = 0 = \frac{T}{S+1} - F_{Rfront} - \frac{F_{Rrear}}{S+1} - D$$

$$0 = \frac{676}{S+1} - F_{Rfront} - \frac{F_{Rrear}}{S+1} - 435$$

where F_R is rolling resistance force, front and rear respectively, and S is the slip ratio (only on the rear, since the Corvette is rear wheel drive).

Before we can proceed, we need to estimate the slip ratio, S. At two-passenger load the Corvette CG is nearly mid-wheelbase. Since the gross weight with two passengers is 3540 lb. the drive wheels will have a total load of 1770 lb. Furthermore, with equal loading the rolling resistance of the front and rear wheels will be about equal. The slip ratio, S, is a function of the tractive force divided by load ($676/1770 = 0.38$) and the slip ratio curve which is nearly linear up to the peak at, say, $S = 0.15$, see RCVD Figure 2.16, p.38. A linear (very rough) estimate of the slip ratio is $S = 0.38 \times 0.15 = 0.057$. Thus,

$$0 = \frac{676}{0.057+1} - \frac{F_{Rtotal}}{2} - \frac{F_{Rtotal}}{2(0.057+1)} - 435$$

$$F_{Rtotal} = 210 \text{ lb.}$$

So, the question now boils down to deciding if 210 lb. of rolling resistance is a reasonable number. Using these numbers, the rolling resistance coefficient is $211/3540 = 0.06$, which is considerably higher than typical numbers measured at low speed (which are about 0.02 for a "sticky" tire like this design).

This puts our analysis in question, but we also know that rolling resistance increases with speed. We checked with the tire manufacturer, Goodyear, and they indicated that there was no measured rolling resistance data for these tires at high speed. Based on their experience, however, they did not consider a value as high as 0.06 to be unreasonable.

There is also a possibility of error in the air drag numbers. The published (minimum) C_D was almost certainly measured in a wind tunnel without the wheels rotating. Rotating the wheels would increase the air drag and reduce the contribution of the rolling resistance.

15.4 Comments on Simple Measurements and Experiments

1. Tufts that lie flat on the surface indicate attached flow, tufts that flutter or point forward indicate stalled flow or turbulence. RCVD Figure 3.12, p.104, shows tufts applied to a wing in a wind tunnel. Note the wingtip vortex and the stalled flow on top of the flap.

2. Almost any material can be (and has been) used to make rapid changes in body shape. Tape, of various kinds, is the universal method for attachment. Modeling clay is normally used in wind tunnels since it can be easily modified yet is not subject to deformation by the airstream.

CHAPTER **16**

Ride and Roll Rates

Solutions to this chapter's problems begin on page 181.

16.1 Problems

1. A vehicle with a 5 ft. front track and 150 lb./in. front wheel center rate (per wheel) is desired to have a front roll stiffness of 1000 lb.-ft./deg. What auxiliary front roll stiffness (anti-roll bar) is required? Assume the tires have an infinite spring rate.

2. Suppose you have done a number of ride rate and roll rate calculations having ignored the tire spring rate. Upon implementation in the vehicle, will your measured chassis-to-ground ride and roll rates be higher or lower than your calculations?

3. Consider a symmetric, 2000 lb. vehicle with 50/50 weight distribution, 8.5 ft. wheelbase, 5 ft. front/rear tracks, CG 1.5 ft. above the ground and independent front/rear suspensions with roll centers on the ground. Determine the front spring rates, rear spring rates and the front anti-roll bar stiffness that fulfill the following criteria:

    - Front ride natural frequency = 1 Hz

- Rear ride natural frequency 10% higher than the front
- Roll gradient = 3 deg./g

This vehicle does not have any provision for a rear anti-roll bar. For simplicity ignore tire spring rates. What roll stiffness distribution results from your calculated rates?

4. It is common to make adjustments during a race by adjusting tire pressures. Consider the quarter car model (single wheel) and suppose that a 1 psi increase in tire pressure increases the tire spring rate 50 lb./in. What is the resultant change in total spring rate (chassis to ground) from a 1 psi increase? What change in wheel center rate would produce the same result? Based on these results, do you think air pressure changes alter vehicle handling through the change in spring rate? Assume nominal wheel center rate and tire rate values of 150 and 1500 lb./in., respectively.

5. Based on RCVD Figure 16.2, p.595 (repeated as Figure 16.1 below), derive the relationship between the wheel center rate K_W and the spring rate K_S in terms of lengths a and b. Identify the installation ratio.

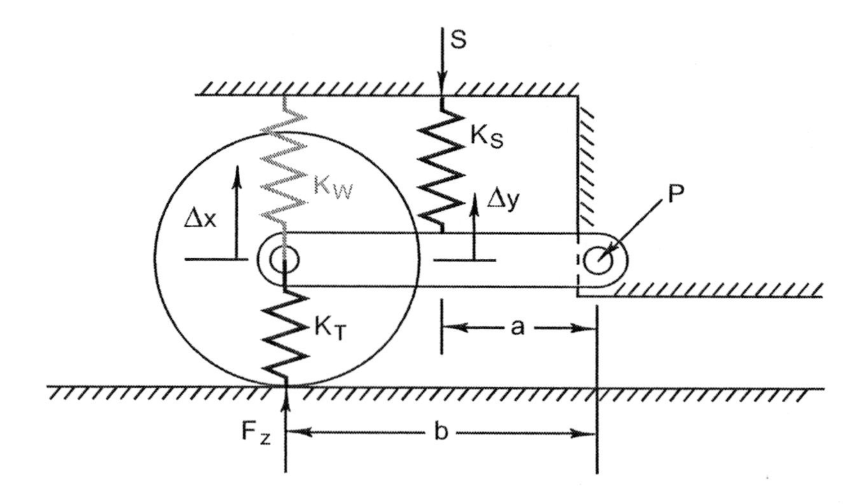

Figure 16.1 *Installation ratio for a simple suspension*

6. A number of different geometric arrangements can be used to produce a rising-rate suspension with a linear spring. The method that seems to be most popular at this time is to attach the chassis end of a push-rod to one arm of a bell crank with an acute angle between the push-rod and the bell crank arm. As the wheel moves up relative to the chassis (bump), the bell crank rotates and more and more of the push-rod load is carried on the pivot of the bell crank (and less load is taken out on the

spring-damper unit that is driven by the other arm of the bell crank). The installation ratio may be determined by calculating the motion ratio between the push-rod and the spring-damper unit for very small deflections.

Figure 16.2 shows a rising-rate push-rod suspension and an approximately linear-rate push-rod suspension as used on a circa 1997 CART or F1 car. Derive an equation that expresses the installation ratio for this suspension across a variety of ride positions. Assume the bell crank, push-rod and spring-damper unit are all in the same plane which is tilted relative to the centerplane of the car. Introduce your own dimensions and make a plot of installation ratio against ride position. To get a feel for the proportions that are currently "in vogue", see *Racecar Engineering*, *Race Tech* or another technical motor racing magazine.

7. A steeply rising-rate installation ratio has been used on cars with high aerodynamic downforce to control ride height (and the air gap under the car). In general terms, what is the effect of vehicle pitch on roll stiffness distribution with this arrangement?

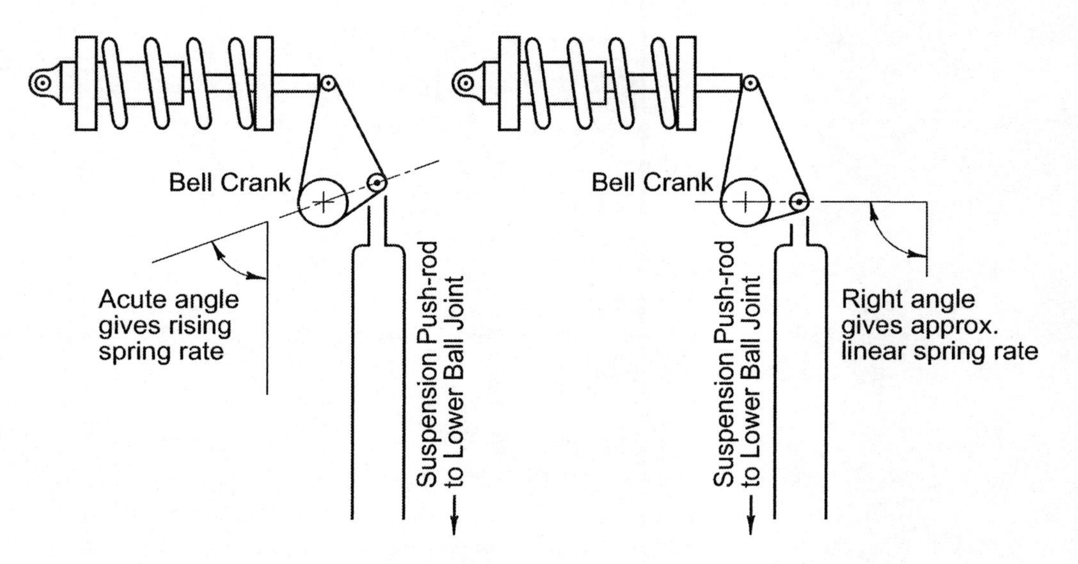

Figure 16.2 *Push-rod/bell crank arrangement: rising-rate (L) and linear (R)*

16.2 Simple Measurements and Experiments

WARNING: The following experiments involve potentially large forces and torques. Do not get underneath a vehicle unless it is supported on jack stands and you are sure that it cannot fall or become unbalanced in any way. Experiment 2 can do significant damage to your vehicle if done on anything but a very smooth surface. Use good judgment!

1. Measure the front and rear total ride rates (lb./in.) for your personal road car. Then, measure the roll stiffness of the car (lb.-ft./deg.). How will you make these measurements? Do you need to do these measurements dynamically?

2. Replace the shock absorbers on one axle of a car with rigid steel bars. It may take some mechanical ingenuity to construct the replacement bars. In a steady-state constant radius test, examine the effects of this modification. Was it what you expected? Suppose the bars were instead placed on the other axle; now what would you expect and what do you observe?

   NOTE: Do not attempt this experiment with vehicles equipped with strut-type suspension systems unless you have the proper spring compression tools. In this case, consider another way of achieving high roll stiffness on one. For example, statically load the car so that it is on its bump stops, then tie the suspension in that position with cables.

   SPECIAL NOTE: Avoid driving over bumps with rigid shock replacements installed. Bump loads in the shock mounts may damage the structure of the car.

3. Remove the ride springs from the car and rate them (see Chapter 21 and/or RCVD Chapter 21). Using the installation ratio, calculate a theoretical ride stiffness for your car based on the spring rate. Why would you expect this to differ from the measurements made in Experiment 1?

4. Remove the anti-roll bar(s) from a car and determine the torsional spring rate of the bar by calculation and then by test (see Chapter 21 and/or RCVD Chapter 21). Determine the expected roll stiffness of the bar(s) by using the torsional spring constant of the bar and the installation geometry. Next, using the ride spring rate and installation ratio information from Experiment 3, calculate a theoretical roll rate for the ride springs. Combine the two (bars and springs) to calculate the car's total roll rate.

16.3 Problem Answers

1. The expression for roll stiffness on an axle is:

$$K_\phi = \frac{1}{2} K_W t^2 + K_{\phi aux}$$

Substituting values:

$$1000 \, \frac{\text{lb.-ft.}}{\text{deg.}} = \frac{1}{2} (150 \times 12 \text{ lb./ft.}) \, (5 \text{ ft.})^2 \left(\frac{\pi \text{ rad.}}{180 \text{ deg.}} \right) + K_{\phi aux}$$

$$K_{\phi aux} = 607.3 \text{ lb.-ft./deg.}$$

2. The measured rates will be lower than the calculations. The presence of a tire spring rate softens the system because the tire rate is in series with the wheel center ride rate and the anti-roll bar.

3. Begin with the front ride frequency. According to RCVD Figure 16.1, p.583, a 1.0 Hz (60 cpm) ride frequency corresponds to a static deflection of 9.8 in. Since each front wheel carries 500 lb., each front spring rate should be:

$$K_{LF} = K_{RF} = \frac{500 \text{ lb.}}{9.8 \text{ in.}} = 51.0 \text{ lb./in.} = 612.0 \text{ lb./ft.}$$

The rear ride frequency is 1.1 Hz (66 cpm), which is 10% higher than the front. This results from a static deflection of 8.1 in., from which the rear spring rates are:

$$K_{LR} = K_{RR} = \frac{500 \text{ lb.}}{8.1 \text{ in.}} = 61.7 \text{ lb./in.} = 740.7 \text{ lb./ft.}$$

Now determine the required roll stiffness. The desired roll gradient is 3 deg./g. A lateral acceleration of 1g results from 2000 lb. of lateral force. This force is acting 1.5 ft. above the roll axis, creating a moment of 3000 lb.-ft. For a 3000 lb.-ft. roll moment to result in 3 deg. of roll, a roll stiffness of 1000 lb.-ft./deg. is needed.

The roll stiffness from the front and rear ride springs can be supplemented by a front anti-roll bar. The roll stiffness provided by the front and rear springs, respectively:

$$K_{\phi F} = \frac{1}{2}(2 \times 612)\left(5^2\right)\left(\frac{\pi}{180}\right) = 267 \text{ lb.-ft./deg.}$$

$$K_{\phi R} = \frac{1}{2}(2 \times 740.7)\left(5^2\right)\left(\frac{\pi}{180}\right) = 323 \text{ lb.-ft./deg.}$$

These sum to 590, leaving 410 lb.-ft./deg. to be taken by the front anti-roll bar. The roll stiffness distribution is:

$$\frac{267+410}{1000} \times 100\% = 67.7\% \text{ front}$$

Note that these calculations do not imply anything about the handling characteristics of the vehicle, nor do they imply anything about occupant comfort.

4. The nominal ride rate, chassis to ground, is (see RCVD p.591):

$$K_R = \frac{K_T K_W}{K_T + K_W} = \frac{(1500)(150)}{1500+150} = 136.36 \text{ lb./in.}$$

Increasing the tire pressure by 1 psi results in a new tire rate of 1550 lb./in. This gives:

$$K_R = \frac{(1550)(150)}{1550+150} = 136.76 \text{ lb./in.}$$

A 50 lb./in. increase in tire rate results in a mere 0.4 lb./in. increase in the ride rate. To get this new ride rate by changing the wheel center rate (instead of the tire rate):

$$136.76 = \frac{(1500)K_W}{1500+K_W} \quad \Rightarrow \quad K_W = 150.5 \text{ lb./in.}$$

Thus, the 50 lb./in. increase in tire rate has the same effect as a 0.5 lb./in. increase in wheel center rate. From these calculations we can conclude that the air pressure effect on spring rate is too small to have a significant effect on vehicle handling. Air pressure adjustments also modify:

- Tire footprint and tread unit loading
- Tire cornering stiffness

- Static wheel loads via diagonal weight jacking
- Tire heating characteristics

It is difficult to quantify these effects.

5. The spring in Figure 16.1 has stiffness K_S and the wheel center rate is denoted by K_W. Taking moments about P gives $F_Z b = F_S a$. Substituting $F_Z = K_W \Delta x$ and $F_S = K_S \Delta y$ into this expression gives:

$$bK_W \Delta x = aK_S \Delta y$$

$$K_W = \frac{a}{b}\frac{\Delta y}{\Delta x} K_S$$

By similar triangles, $\Delta y / \Delta x = a/b$ and so:

$$K_W = \left(\frac{a}{b}\right)^2 K_S$$

where the installation ratio, $IR = a/b$.

6. Figure 16.3 establishes the nomenclature for this calculation. It is drawn at static ride position, $\phi = 0$.

Taking moments about the bell crank pivot, P:

$$\ell_1 F_1 \cos(\theta_1 + \phi) = \ell_2 F_2 \cos(\theta_2 + \phi)$$

The forces are related to the spring rates by:

$$F_1 = K_W \Delta_1 = K_W \ell_1 \sin\phi \cos(\theta_1 + \phi)$$

$$F_2 = K_S \Delta_2 = K_S \ell_2 \sin\phi \cos(\theta_2 + \phi)$$

Substituting into the original expression gives:

$$\ell_1^2 K_W \sin\phi \cos^2(\theta_1 + \phi) = \ell_2^2 K_S \sin\phi \cos^2(\theta_2 + \phi)$$

$$\frac{K_W}{K_S} = \left(\frac{\ell_2}{\ell_1}\right)^2 \left(\frac{\cos^2(\theta_2 + \phi)}{\cos^2(\theta_1 + \phi)}\right)$$

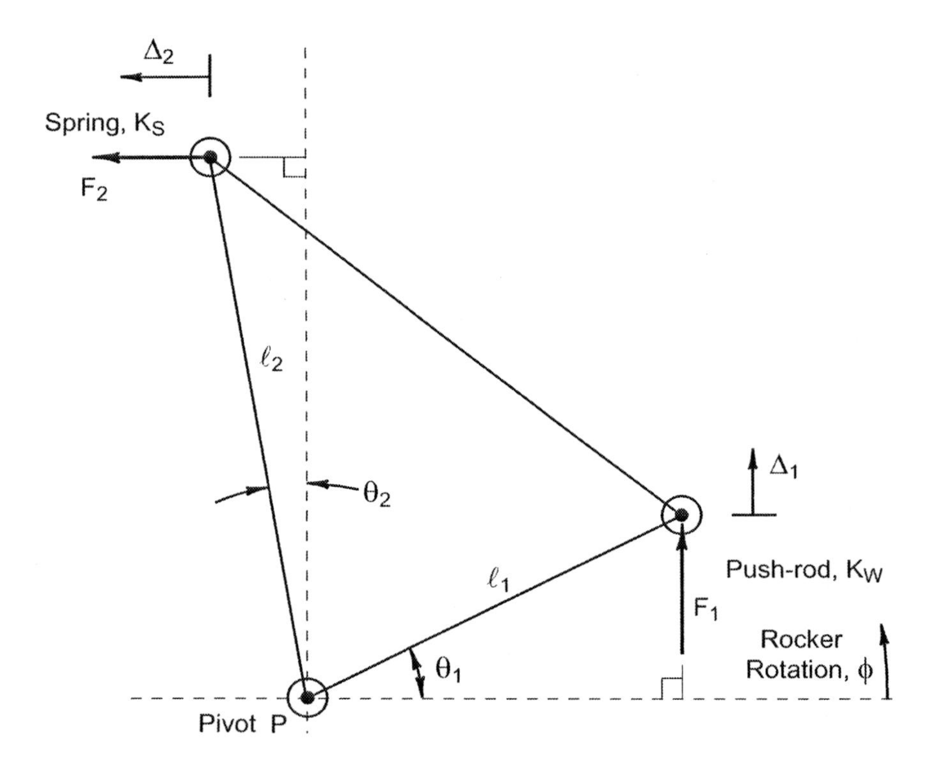

Figure 16.3 *Schematic for bell crank calculations*

When $\theta_1 = \theta_2$ this equation reduces to the standard installation ratio formula. Figure 16.4 plots installation ratio against ride position for several bell crank designs.

7. Rising-rate installations cause both the ride stiffness and roll stiffness to be a function of wheel ride position. If the car is pitched nose-down, the front suspension is compressed and operates at a relatively high spring rate. This results in high front roll stiffness. At the same time the rear suspension is extended and operating at a relatively low spring rate (and, therefore, low roll stiffness). The reverse is true if the car is pitched nose-up.

16.4 Comments on Simple Measurements and Experiments

1. One way to measure overall ride rates is through static loading of the vehicle with known weights, measuring the deflection at each load. It is very important to make measurements when loading and also when unloading to account for suspension friction. Use all the data points in a regression to determine the ride rate.

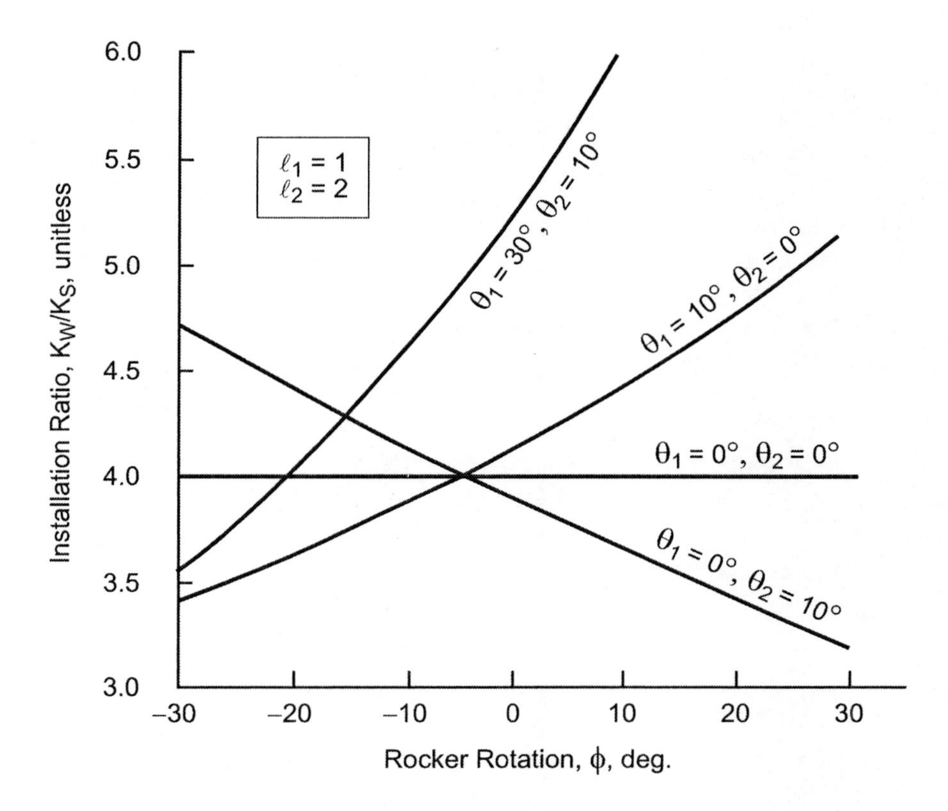

Figure 16.4 *Variation of installation ratio with ride position*

A variation of the method given above avoids the need to measure twice (loading and unloading) by removing the suspension friction effects during the test. This can be done by shaking (dithering) the car manually at each new load (and ride position) to settle the car in the middle of the friction deadband. In this case the accuracy of the test will depend on the friction level in the suspension and the judgment of the person performing the test.

An alternate method of determining the ride rate is to remove the suspension dampers and oscillate the car vertically to determine the natural frequency in ride. With that information, the chart (and formula) given in RCVD Figure 16.1, p.583, can be used to determine the static deflection. The static deflection and the static wheel loads can be used to calculate the ride rate. Caution must be used with this method because friction in the suspension will make the apparent natural frequency amplitude-dependent. At low amplitudes the suspension may not move and the tire spring rate will dominate the measurement. This problem is particularly noticeable with multi-leaf suspensions (high friction) as discussed by Olley[1].

---

[1] See Section 5.2, p.272, of *Chassis Design: Principles and Analysis*, Milliken/Milliken, SAE R-206, 2002.

Overall roll rates are more difficult to measure. Olley proposes a technique using a constant amount of ballast that is shifted from one side of the car to the other[2]. Again, the suspension will have friction that must be taken into account when reducing the data.

2. When the ride springs on a street car are replaced with solid bars, the ride rate and roll rate are increased dramatically. The tires become the primary springs in the system and they are very stiff. The increase in roll stiffness should degrade the lateral force capability of the pair of tires on that end of the car. If the rigid links are on the front the tendency will be toward "push" and if they are on the rear the tendency will be toward "spin".

3. Measured and calculated ride rates will rarely agree for a variety of reasons. For example:

   - The installation ratio is difficult to measure on a car.
   - The tire spring rates were not measured or taken into account in the problem statement.
   - Friction in the overall ride rate measurements introduces error.

4. For packaging reasons, many anti-roll bars are not simple straight torsion springs. Calculating the spring rate may require advanced techniques such as finite element analysis. A spring rate test (with the bar constrained as it is constrained in the car) will usually be the quickest and most accurate method to rate a bar of this type. As with Experiment 3, there are a variety of potential problems with simple calculations of roll rate.

[2]Ibid., Section 8.8, p.469.

CHAPTER **17**

Suspension Geometry

Solutions to this chapter's problems begin on page 193.

17.1 Problems

The first four problems in this section are considerably more involved than the other problems in this book. Successful solution requires a good working knowledge of engineering mechanics. We begin this series of problems with some introductory material.

Bill Shope, formerly of the Chrysler Vehicle Performance Department, suggested that we include the equations required to tune the three-link and Panhard bar rear axle suspension for equal rear wheel loads on acceleration. This suspension arrangement is shown in RCVD Figure 17.38, p.653. As we investigated this problem we found that it is not only difficult and interesting but illustrates a number of significant aspects of vehicle dynamics.

First, some background from a draft article by Shope entitled, "The Benefits of Equality", on the advantages of equal rear wheel loading for dragsters:

> The car is safer because equal rear tire loading translates into equal rear tire thrust meaning that, when the front tires leave the track surface, the car will not veer to the right or left.

The car performs better because of the characteristics of the pneumatic tire; whether in traction or in cornering, the total force available is greatest when the tires are equally loaded.

With a conventional driveline, however, the driveshaft is acting to unload the right rear and load the left rear. But, since every force (or torque) has an equal and opposite reaction, there must be a torque available to counteract the effect of the driveshaft torque. That torque is transmitted into the chassis through the engine and transmission mounts. From there, it is distributed to the front and rear suspensions in direct proportion to the relative roll stiffness.[1]

Conventional solutions to this problem include:

- Softening the front suspension in ride also softens it in roll, so this will reduce the portion of the torque reaction taken on the front relative to the rear. This also increases the CG height rise on acceleration (more rearward weight transfer) but Shope feels that the more-equal rear wheel loading is the significant advantage to soft front springs.
- Stiffening the rear in roll (increasing the proportion of the roll torque taken at the rear) could also be done with stiff rear springs and a rear anti-roll bar, but this will increase the changes in rear tire loading from drag strip surface irregularities.

    If the reaction torque must be transmitted to the rear axle through suspension springs, the oscillation frequency (for body roll) will be low enough that tire loadings will be directly affected. Left and right rear tire loadings will oscillate about their final values, through an appreciable range, for about two seconds after launch. In addition, weight transfer to the rear wheels also will vary about its final value at a similar frequency. This is clearly seen on a car which exhibits a large amount of squat or rise on acceleration. Such a car will overshoot its final attitude before settling. This visible movement corresponds to changes in total wheel loading.

    So, even if all the reaction torque is returned to the rear axle, the presence of suspension springs will cause tire load oscillations that adversely affect performance and safety. Note that this is true even with a perfectly adjusted four link rear suspension. Reaction torque is still transmitted through the suspension springs.

The dragster solution, solid mounting of the rear axle to the chassis, is effective only if the connection from the rear axle to the engine/transmission is torsionally rigid.

---

[1] See also Section 18.9 in *Race Car Vehicle Dynamics*.

Shope's interest in the offset three-link derived from the rear suspension on the C-type Jaguar of the 1950s. He applied this technology to the famous Ramchargers drag race car but only after he developed a comprehensive analysis for the link locations and angles. The benefit of an analysis was illustrated in the 1959 NHRA Nationals when the three-link was not properly adjusted and it overcompensated, bouncing both right side wheels off the track!

Bill Shope has kindly provided us with his derivation of the mechanics of a specific offset three-link (plus Panhard bar) rear axle suspension. His solution gives the position and angle of the offset link and the angularity of the other two links, for the special case where the left and right rear wheel loads remain equal on acceleration. His suspension geometry also provides 100% anti-squat to insure that the link angles do not change during acceleration (which would change the equal wheel loading). A feature of this axle suspension is that the rear wheel loads are equalized for any level of driveshaft torque, provided that the link angles/locations are chosen to match the rear axle ratio used.

Norman Smith, a former employee of Jaguar in the suspension design and handling area has independently derived equations for an offset three-link suspension as used on the C-type Jaguar. This analysis does not include the 100% anti-squat feature. Incidentally, Smith provides some historical background for this suspension development at Jaguar. With the small tires of that era and lacking limited-slip differentials, Jaguar's competition cars were limited in acceleration. The offset three-link arrangement was the work of Bill Heynes, Bob Knight and Jim Randle.

1. Consider three-link axle suspension of Figure 17.1. Develop the equations for the lateral location and slope of the laterally displaced upper link which gives equal loads on the two wheels under acceleration (for a given final drive axle ratio). Assume straight line operation and hence no lateral forces or body roll due to lateral acceleration or drive torque (which is completely compensated in the suspension). Ignore any provision for anti-squat in the suspension. Also develop the equation(s) for the slope of the lower links (same for both).

2. Using the equations developed in Problem 1, determine the slopes of the single upper and two lower links for rear axle gear ratios G = 3 and G = 4. Use the following input data:

   - $y = 30$ in.
   - $H = 20$ in.
   - $d_U = 23$ in.
   - $d_L = 7$ in.
   - $R = 13$ in.

   Draw a side view elevation for each case. Some adjustments in the location of the side view swing arm length (B) may be required.

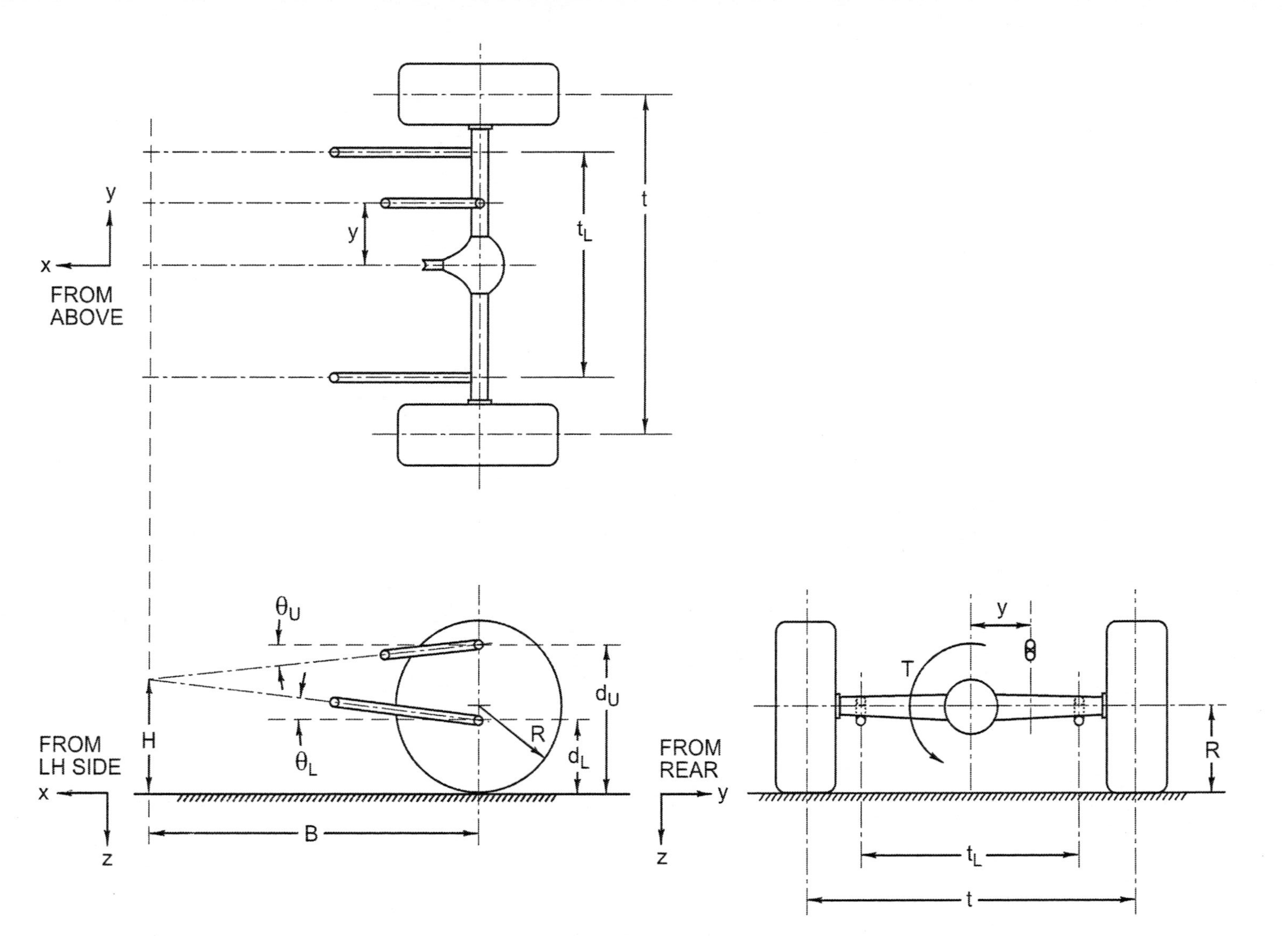

***Figure 17.1** Three-link schematic layout*

3. Develop equations as described in Problem 1, but with the additional condition of 100% anti-squat in the suspension.

   Hint: In developing the free-body diagrams for his solution, Shope used horizontal and vertical components of the forces in the three links, thus avoiding much trigonometry.

4. Using the equations from Problem 3, determine the slopes and offsets of the links for the input data of Problem 2. Additionally:

   - Rear axle ratio $G = 4$
   - Vehicle wheelbase $\ell = 100$ in.

   Draw a side view elevation for the 100% anti-squat condition.

5. The description of jacking in RCVD Section 17.3 pp.613-15, discusses the forces on the outer wheel only. In reality, the inner wheel also has horizontal-vertical force coupling. Redraw RCVD Figure 17.8 to include both wheels on an axle. What happens to the height of the CG when equal lateral forces are applied to each wheel? What happens when unequal lateral forces are applied (as is usually the case with weight transfer in a turn)?

   Replace the swing axle in the figure with an SLA suspension. What happens to the height of the CG when lateral force is applied at the tires with the roll center above ground level? What if the roll center is below ground?

6. [2]Consider the effects of axle roll steer or, really, any form of suspension roll steer. This is briefly treated in RCVD Section 17.4, p.622, and is treated in much greater detail in *Chassis Design: Principles and Analysis*[3]. What would you expect to happen to the tire slip angles and the vehicle attitude angle in a circular skidpad test when the rear roll steer is changed? What is the effect on steering wheel angle and the linear range over/understeer (if any)? What is the effect on final plow/spin characteristics?

7. [4]What is the effective difference (if any) between the three types of Watt's links shown in Figure 17.2? The first question is whether the center link should be attached to the body (as in a) or the axle (as in b). Consider two cases, a road-racing car with very limited suspension travel and a desert racer with a great deal of ride travel. Then discuss any effects from rotating the linkage into a horizontal plane.

---

[2]This problem requested by Bill Shope, formerly of the Chrysler Vehicle Performance Department.

[3]Milliken & Milliken, *Chassis Design: Principles and Analysis*, SAE R-206, 2002.

[4]This question came from Richard Welty who owns an Alfa-Romeo with the center link attached to the axle.

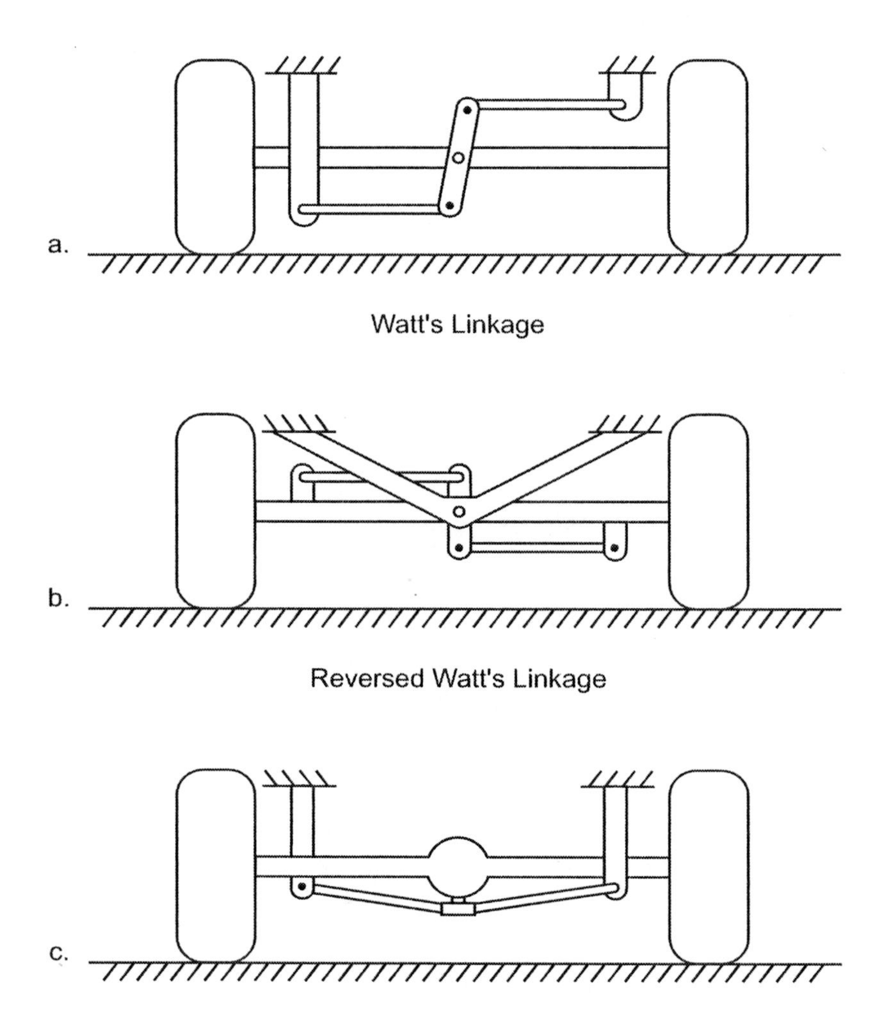

Figure 17.2 *Three types of Watt's link*

17.2 Simple Measurements and Experiments

1. With assistants to stand on the front bumper (as variable ballast), measure the camber changes that occur during compression of the front suspension of a car. The camber measurements will require a camber gauge or a precision inclinometer, often available in machine shops. What can you deduce about the suspension geometry and the structure of the car from these measurements?

17.3 Problem Answers

1. As noted in the introduction to the problems for this chapter, the idea for the three-link problems came from Bill Shope. We also communicated with Norman Smith (ex-Jaguar) and we give the full text of his letter/analysis below as the answer to Problem 1 (three-link without 100% anti-squat). Square brackets are used for a few editorial notes.

   I. The Problem

   The C-type Jaguar had a high acceleration potential (for its day), but a live rear axle. This was subject to torque about the propeller shaft axis which transferred load from one driving wheel to the other, even under straight line acceleration conditions. This limited the acceleration available in the lower gears, as the lightly loaded right-hand wheel would spin before the more heavily loaded left-hand wheel. The open differential ensured that the torque applied to the two wheels was more-or-less equal, resulting in a left-hand tyre whose friction was not fully used.

   The solution was either to install an independent rear suspension, where the propshaft torque is resisted, not by the tyres, but by the final-drive mountings, or to devise some suspension modifications that redistribute the wheel vertical loads to be equal. The commonest such modification is the torque tube, but Jaguar chose to use a single offset link, to which the remainder of this description is confined.

   II. The C-type Rear Suspension

   • Initial Description

   The Jaguar C-type[5] describes the C-type suspension as:

   "The rear suspension was, however, quite original, being via a single transverse torsion bar passing through the lower tubular chassis frame member and anchored at the centre. A trailing link from each end of the torsion bar was connected to hanger brackets attached under the rear axle casing. The primary locating member was an upper link, or A-bracket, just inboard from the offside rear wheel [right hand rear wheel]. This not only located the axle laterally, but was designed to work in reaction to the torque produced by hard acceleration, reducing the tendency for the right-hand wheel

[5] *The Jaguar C-type*, Profile Publications number 36 (12 pgs.), pub. 1966 by Profile Publications Ltd. at 2 shillings (US & Canada – 50 cents).

to lift. Thus, although an orthodox axle was used, road grip proved as effective as that provided by many contemporary de Dion systems."

It is perhaps important to realize that the trailing links would provide little lateral location as they were thin horizontally and would have to bend considerably before generating any lateral forces.

Figure 17.3, taken from *Jaguar Sports Racing Cars*[6], is a reproduction of Bob Knight's sketch of the initial design concept for the C-type Jaguar suspension, which describes it as follows:

"The rear suspension was completely different from the road car (XK120). A Salisbury axle with offset hypoid bevel final drive was attached to the rear of the chassis frame by trailing links.

"On the right-hand side an 'A' bracket, or reaction plate, was fitted above the axle housing and connected to the rear upright of the chassis. This was designed to have a dual purpose. The first was to provide solid lateral location of the axle within the car. The second was to avoid and harness the rotational forces exerted to the rear axle under heavy acceleration. In fact this reaction member translated the forces into downward pressure on the axle, improving adhesion. Additionally the tendency for weight to be transferred from the offside [right hand] to the nearside [left hand] rear wheel during fierce acceleration from rest or low speed was avoided by positioning the reaction member a specific distance from the centre of the axle, the downward pressure thus equaling the weight transfer."

- History

While the above suspension improved traction for straight-line acceleration, the rear roll centre was 16 inches above the ground at the height of the top link. It was recognized that this gave insufficient understeer both in a straight line and in corners. The top A-link was replaced by a radius rod lying substantially fore-aft which functioned in the same manner for acceleration as the A-link. Lateral location was now performed by a Panhard rod, which crossed the car centre-line underneath the axle, lowering the rear roll centre by 6 inches.

This formed the basis or the "production" C-type rear suspension and is shown in Figure 17.4 which comes from *Jaguar Sports Racing Cars*. The transverse shaft [shown in the figure] on top of the axle is probably the eas-

[6]*Jaguar Sports Racing Cars* by Philip Porter, pub. 1995 by Bay View Books, Ltd., The Red House, 25-26 Bridgeland Street, Bideford, Devon EX39 2PZ, UK. ISBN 1 870979 67 2.

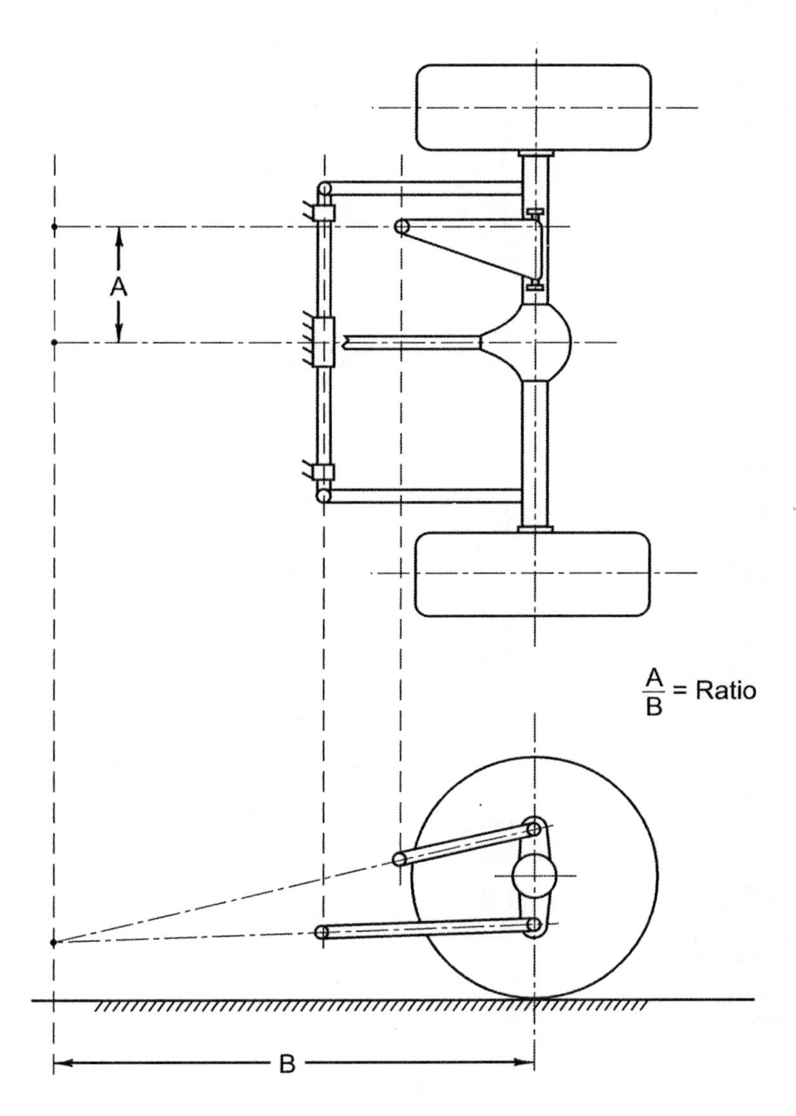

Figure 17.3 *Original conceptual sketch of C-type suspension by Bob Knight*

iest way of connecting the radius rod to the existing pick-up points on the axle designed for the earlier A-link.

The very last, lightweight C-types had rear suspensions which were essentially prototypes for the rear suspension of the D-type of 1954. This suspension was symmetrical with no torque compensation, possibly because the rear axle was now fitted with a Powr-Lok limited slip differential which would, to some extent, allow torque to be biased to the wheel reacting the

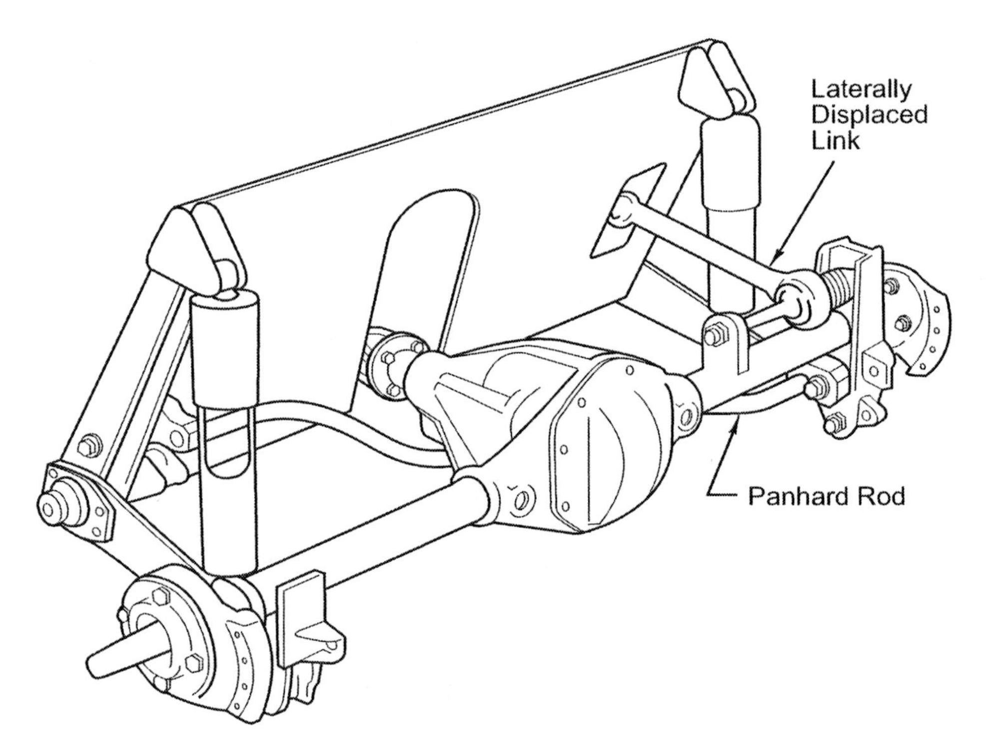

Figure 17.4 *"Production" C-type rear suspension*

greater vertical force.

III. Analysis

• Preamble

The torque developed between the body and the rear axle, generated by the engine-gearbox and transmitted by the propeller shaft on one end and the engine mounts on the other, results in body roll once transients have taken place. If no torque tube or special linkage is used, this roll is proportional to the torque and inversely proportional to the total suspension roll stiffness. This steady-state effect is analyzed in RCVD, pp.697-700.

It is worth noting that the distribution of roll stiffness front-to-rear determines the steady-state vertical force imbalance on the rear axle. If all the roll stiffness were apportioned to the rear, the rear springs would carry unbalanced forces which would exactly oppose the propeller shaft torque in the steady-state after transients. This is unlikely to be the case due to the handling implications of such a roll stiffness balance.

If a suitable suspension link, or some other modification, compensates for the propeller shaft torque by rechanneling the forward thrust forces to rebalance the tyre vertical forces, no roll moment is transmitted to the body, no roll takes place and therefore there are no [roll] transients.

- Force Systems to be Analyzed

The analysis below considers the system of forces and moments acting on the rear axle from the body (through the suspension links), from the propeller shaft and from the ground. The axle is assumed to be in equilibrium, as the mass and acceleration of the axle are ignored.

No lateral forces are considered, since the only agent generating such forces would be the A-link (or in later cars the Panhard rod), and their presence would imply lateral reactions at the tyres, which could only be present in other than straight line conditions.

The forces due to squat of the body acting through the springs are also ignored, since, for symmetrical squat, the effects on the tire-ground forces are symmetrical and thus provide no compensation for propeller shaft torque. Further, if the anti-squat properties of the suspension linkage are appropriate, there may be no squat at the rear suspension at all (see Problems 3 and 4).

Body roll is also ignored, because in straight-line acceleration there will be no roll due to lateral forces. It is assumed that the linkage to be investigated is providing 100% torque compensation, so that the torque acting on the body through the engine mounts is exactly balanced by the rear suspension linkage torque derived from the forward thrust forces.

- Analysis of the Forces due to Transmission Torque Only

In the following, F is used to represent a force, with subscripts U, L indicating the top [upper] link and bottom [lower] links, respectively. Additional subscripts X and Z indicate force components in the longitudinal and vertical directions, and L & R apply to forces split between left and right sides of the car.

The force in the top link acts along the link, since it is pivoted at both ends. The lower link force acts along the links because they are pivoted to the axle. Since no spring forces are considered here, no torque from the torsion bar ride spring acts at the forward end of the links.

Taking moments about the lower link connection to the axle in side view (see Figure 17.5a):

$$F_{UX}(p+q) - F_x(H-q) = 0$$

$$F_{UX} = \frac{H-q}{p+q} F_x$$

In top view (Figure 17.5b), since there is an open differential in the axle, the thrusts from the two wheels must be equal:

$$F_{xL} = F_{xR} = F_x/2$$

So, the moment about the axle centre:

$$F_{LXL}\left(\frac{t_L}{2}\right) - F_{LXR}\left(\frac{t_L}{2}\right) - F_{UX}y$$

$$F_{LXL} - F_{LXR} = \left(\frac{2y}{t_L}\right) F_{UX}$$

In rear view (Figure 17.5c), for 100% torque compensation $F_{zL} = F_{zR}$. Denote the propeller shaft torque applied to the axle as T.

Then, moments about the axle centre:

$$(F_{LZR} - F_{LZL})\left(\frac{t_L}{2}\right) + F_{UZ}y = T$$

Noting that:

$$F_{LZR} = F_{LXR} \times (q/B)$$

$$F_{LZL} = F_{LZR} \times (q/B)$$

$$F_{UZ} = F_{UX} \times (p/B)$$

gives:

$$(F_{LXR} - F_{LXL})\left(\frac{t_L}{2}\right)\left(\frac{q}{B}\right) + F_{UX}\left(\frac{yp}{B}\right) = T$$

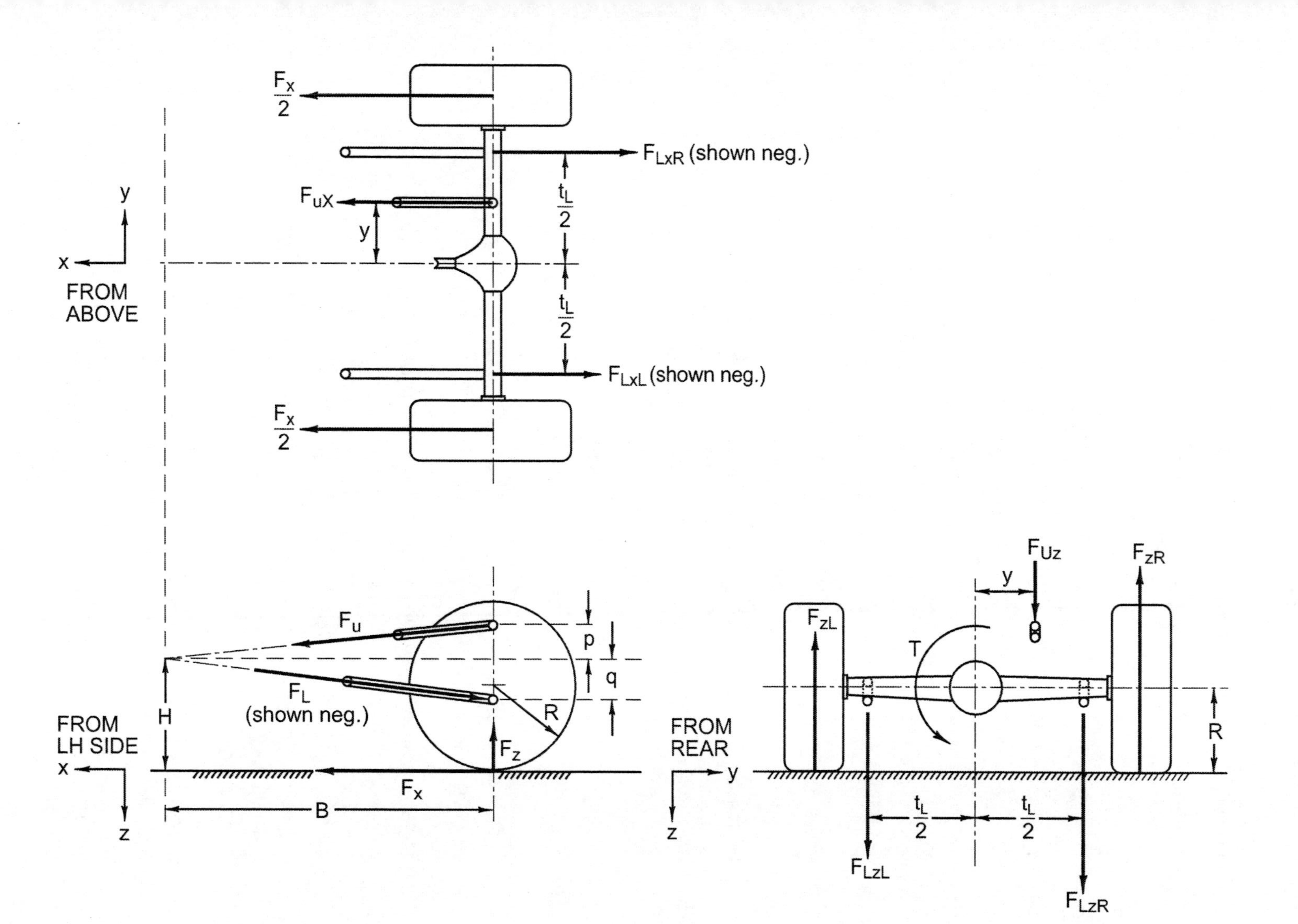

Figure 17.5 *Force/moment free body diagrams*

Via substitution from the top view equation:

$$F_{UX}(p+q)\left(\frac{y}{B}\right)=T$$

And via substitution from the side view equation:

$$F_X(H-q)\left(\frac{y}{B}\right)=T$$

Now, in a simple model of the final drive with gear ratio G:1, tractive effort $F_X = TG/R$. Substitution for F_X in the above equation gives the requirement for 100% torque compensation to give equal tyre vertical loads:

$$\frac{y}{B}=\frac{R}{G(H-q)}$$

From the relationship established above for torque compensation, it is worth comparing Bob Knight's original sketch (Figure 17.3) with the case where the lower links are at hub height and horizontal ($q = 0$ and $H = R$). This reduces the requirement to:

$$\frac{y}{B}=\frac{1}{G}$$

Another case of interest is where $q = H$, if the lower links meet the axle at ground level. In this case, y/B would have to be infinite, as no force would be induced in the top link.

The above analysis could be made more complete if:

- Account were taken of the mechanical losses between propeller shaft torque and forward thrust from the tyres.
- R was taken as the effective rolling radius of the tyre.
- The links are not assumed to be parallel to the car centre-line when seen in plan view.

However, it is probable that the result would not differ substantially from that derived.

2. The previous problem derived the following relationship to give 100% torque compensation and equal tire vertical loads under straight line acceleration:

$$\frac{y}{B} = \frac{R}{G(H-q)}$$

The slope of the upper and lower links, respectively, are given by the following equations. The slope and angle are positive if the link travels upward as it travels forward.

$$\tan\theta_U = -p/B$$

$$\tan\theta_L = q/B$$

where in Figure 17.6:

$$p = d_U - H = 23 - 20 = 3 \text{ in.}$$
$$q = H - d_L = 20 - 7 = 13 \text{ in.}$$

For a final drive ratio, G, of 3.0:

$$\frac{30}{B} = \frac{13}{3(20-13)} \quad \Rightarrow \quad B = 48.5 \text{ in.}$$

$$\theta_U = \tan^{-1}\left(\frac{-p}{B}\right) = \tan^{-1}\left(\frac{-3}{48.5}\right) = -0.0618 \text{ rad.} = -3.54 \text{ deg.}$$

$$\theta_L = \tan^{-1}\left(\frac{q}{B}\right) = \tan^{-1}\left(\frac{13}{48.5}\right) = 0.2621 \text{ rad.} = 15.0 \text{ deg.}$$

For a final drive ratio, G, of 4.0:

$$\frac{30}{B} = \frac{13}{4(20-13)} \quad \Rightarrow \quad B = 64.6 \text{ in.}$$

$$\theta_U = \tan^{-1}\left(\frac{-p}{B}\right) = \tan^{-1}\left(\frac{-3}{64.6}\right) = -0.0464 \text{ rad.} = -2.7 \text{ deg.}$$

$$\theta_L = \tan^{-1}\left(\frac{q}{B}\right) = \tan^{-1}\left(\frac{13}{64.6}\right) = 0.1986 \text{ rad.} = 11.4 \text{ deg.}$$

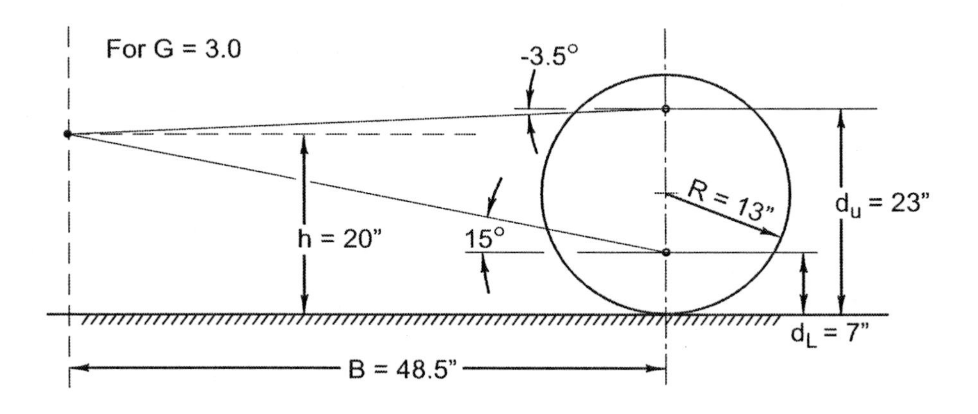

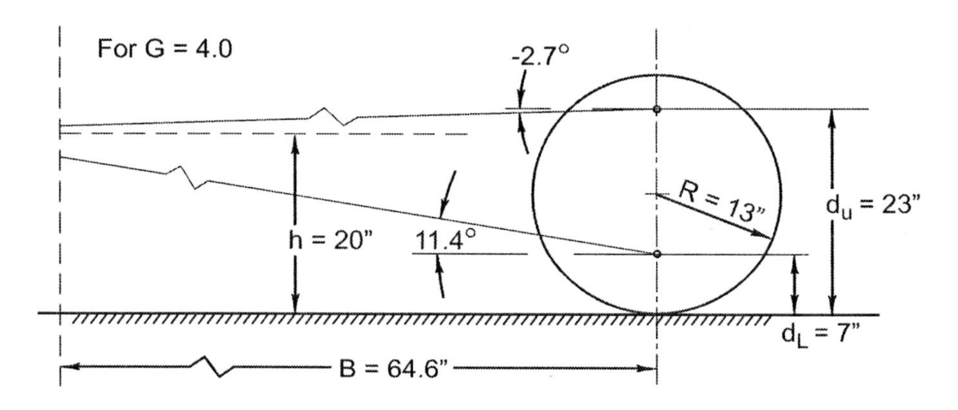

Figure 17.6 *100% torque compensation for different axle ratios*

3. Bill Shope's analysis begins by writing all of the force and moment equations that define the axle and vehicle dynamic equilibrium. By suitable substitution Shope is able to eliminate variables to arrive at final expressions for link angles and locations.

   First some definitions:

h	sprung mass vehicle CG height
ℓ	wheelbase
R_ℓ	tire loaded radius
t	wheel track (to center of tires)
t_L	lateral spacing of lower links at the axle
d_U	distance from ground to rear upper link mounting point
d_L	distance from ground to rear lower links mounting points

y	lateral offset of upper link from centerline of car
F_X	total tire thrust
ΔF_Z	total normal reaction force at the rear wheels
T	driveshaft torque (from transmission)
G	axle ratio (i.e., 4.11 for a 4.11:1 axle)
L_X	horizontal force component of lower left link
L_Z	vertical force component of lower left link
R_X	horizontal force component of lower right link
R_Z	vertical force component of lower right link
U_X	horizontal force component of upper link
U_Z	vertical force component of upper link
θ_U	angle up from horizontal of upper link
θ_L	angle up from horizontal of lower link

Note that ΔF_Z does not include the static weight which is carried by the springs. Note also that Shope has already decomposed the forces in the links into their horizontal and vertical components. Using this notation, free-body equations can be written. The sum of the horizontal forces (in side view):

$$F_X + U_X = L_X + R_X$$

Moments about the left tire ground contact patch in top view:

$$F_X\left(\frac{t}{2}\right) + U_X\left(\frac{t}{2} + y\right) = L_X\left(\frac{t}{2} - \frac{t_L}{2}\right) + R_X\left(\frac{t}{2} + \frac{t_L}{2}\right)$$

$$F_X t + U_X(t + 2y) = L_X(t - t_L) + R_X(t + t_L)$$

The sum of the vertical forces (rear view):

$$\Delta F_Z = U_Z + L_Z + R_Z$$

Moments about the left tire ground contact patch in rear view:

$$\Delta F_Z t + 2T = U_Z(t + 2y) + L_Z(t - t_L) + R_Z(t + t_L)$$

Anti-squat moments about the front tire ground contact point in side view:

$$\Delta F_Z \ell = F_X h$$

Axle housing moments about the rear tire ground contact point in side view:

$$d_U U_X = d_L (L_X + R_X)$$

Side view moment about the axle shafts:

$$F_x R_\ell = TG$$

The lower links are at the same angle to the horizontal in side view:

$$L_Z R_X = R_Z L_X$$

The above force and moment equations can be solved simultaneously to give a general formulation for the angles and locations of the three links. The derivation is not given here—while solution by hand is possible, the use of a computer algebra system simplifies the process.

The upper link tangent in side view:

$$-\tan\theta_U = \frac{U_Z}{U_X} = \frac{d_U R_\ell}{y d_L G} - \frac{h}{\ell}$$

and the lower link tangent (left and right links are at the same angle):

$$\tan\theta_L = \frac{L_Z}{L_X} = \frac{h}{\ell} - \frac{R_\ell}{yG}$$

Subsequent correspondence from Bill Shope examined the case where a single lower link is offset and the two symmetric links are above the axle. He looks at this case for packaging reasons—the offset required for the offset link is less in this case. This becomes important with modern drag race cars where the wide tires and narrow track leave very little room for the rear suspension. Quoting from Bill Shope:

> Reducing the tangent equations to forms involving significant parameter ratios has caused me to realize that a convenient generalization is now available. If the angle is considered positive when the link is pointing up from its rear mounting point in all cases, there is but one equation:

$$\tan\theta = \frac{h}{\ell} - K\frac{R_\ell}{yG}$$

where K has one of three values:

$$K = \begin{cases} 1 & \text{When the angle is associated with the two symmetrically placed links} \\ d_U/d_L & \text{When the angle is associated with an offset upper link} \\ d_L/d_U & \text{When the angle is associated with an offset lower link} \end{cases}$$

These link locations will give equal rear wheel loads on acceleration (on a smooth surface and with a fixed axle ratio, G) and 100% anti-squat geometry. 100% anti-squat effectively decouples longitudinal acceleration from change in rear ride height—no jacking up or down, avoiding any change in wheel loads that would accompany rear ride motion as a function of tractive effort.

Of course, rear wheel load will increase due to longitudinal weight transfer on acceleration. An ideal drag racer would just lift the front wheels but maintain static rear ride height, to avoid exciting rear ride dynamics.

4. The upper link angle in side view:

$$-\tan\theta_U = \frac{U_Z}{U_X} = \frac{d_U R_\ell}{y d_L G} - \frac{h}{\ell}$$

$$-\tan\theta_U = \frac{(23)(13)}{(30)(7)(4)} - \frac{20}{100}$$

$$\theta_U = -0.156 \text{ rad.} = -8.86 \text{ deg.}$$

and the angle for the lower links:

$$\tan\theta_L = \frac{L_Z}{L_X} = \frac{h}{\ell} - \frac{R_\ell}{yG}$$

$$\tan\theta_L = \frac{20}{100} - \frac{13}{(30)(4)}$$

$$\theta_L = 0.0917 \text{ rad.} = 5.24 \text{ deg.}$$

Figure 17.7 shows the resulting suspension layout. The side view swing arm length is B = 64.3 in. and the instant center is H = 12.9 in. above the ground. As a check:

$$\%\text{ anti-squat} = \frac{\tan\theta_R}{h/\ell} \times 100 = \frac{.202}{.2} \times 100 = 101\%$$

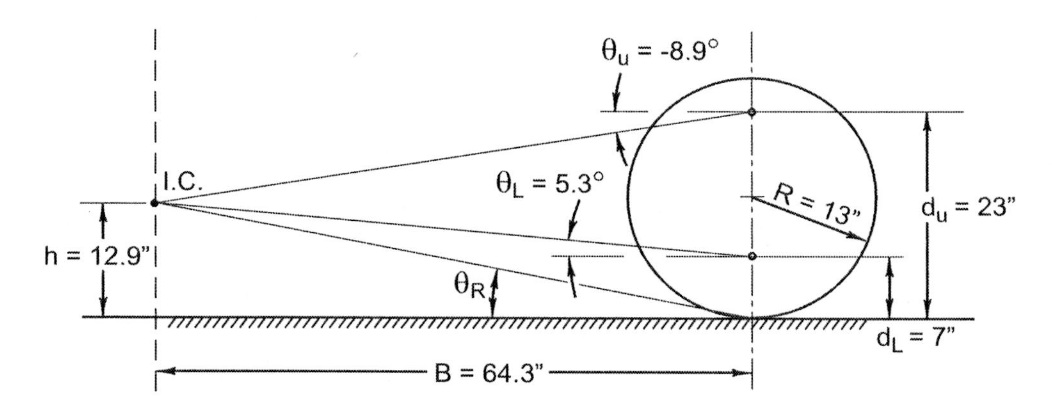

Figure 17.7 100% anti-squat suspension

5. Figure 17.8 shows a pair of wheels with equal lateral force generated by each wheel. The moments about the swing axle instant centers are equal and the CG height does not change significantly. Put another way, the jacking-up moment from the outer wheel is just balanced by the jacking-down moment from the inner wheel.

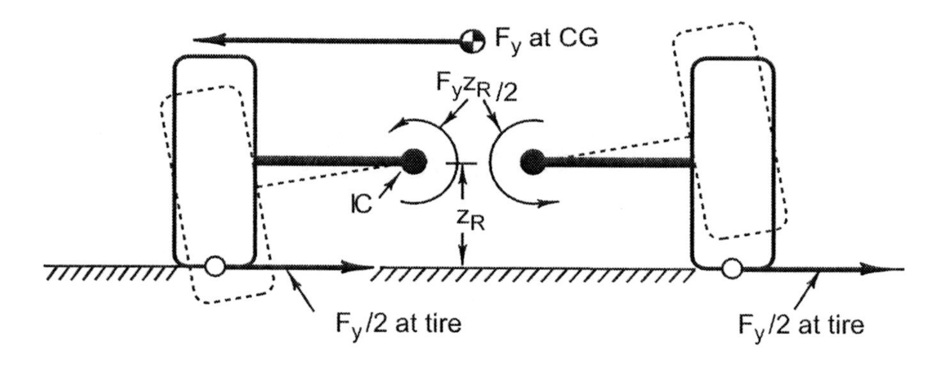

Figure 17.8 Jacking forces on a pair of wheels

With load transfer in a turn, the outer wheel is more heavily loaded and generates more lateral force than the inner wheel. The corresponding jacking-up moment from the outer wheel is larger than the jacking-down moment from the inner wheel and the vehicle CG is raised.

An SLA suspension is illustrated in RCVD Figure 17.7, p.614. It shows both instant centers and the roll center above ground. With weight transfer in a corner and the roll center above ground the car will jack-up in a turn. If the roll center is below ground the car will jack-down.

One could imagine a very asymmetric car with instant center heights above and below ground for the outer and inner wheels, respectively. A calculation of the moments (using tire data to determine the actual lateral forces) may be required to determine if the car jacks-up or jacks-down.

Another interesting jacking case occurs with tire toe-in (or toe-out). In this case, the lateral forces from the two tires will be in opposite directions for straight running and in very low lateral acceleration turns. This gives the possibility of the car jacking-up or down without additional lateral acceleration.

6. Start by assuming that the roll steer is the same on both sides of the car, i.e., if the car is in a right hand turn, it rolls to the left and both wheels on the rear axle steer an equal amount to the right.

   In a circular skidpad test:

   - The individual tire slip angles at any given lateral acceleration (or speed on the skidpad) will not be affected by roll steer.

   - If the steer is as described above (rear axle steer in the same direction as the vehicle is turning) then the effect is understeer. When compared to a vehicle with no axle roll steer, this vehicle will corner in a more "nose out of the turn" attitude, or at a smaller vehicle slip angle, β.

   - Again assuming roll understeer, the car with rear axle roll steer will require a greater steering wheel angle at a given lateral acceleration and it will be more understeer in the linear range.

   - Axle roll steer has no first-order effect on steady-state limit behavior. A change in vehicle slip angle, β, may have some slight secondary effects, especially on very tight radius turns where small angle assumptions are no longer valid.

7. If the center link of the Watt's link is attached to the axle, the roll center stays fixed in the axle and effectively fixed in height above the ground (ignoring the effect of tire deflection). In the "reversed" Watt's link with the center link fixed to the chassis, the roll center is attached to the chassis and moves up and down (relative to the ground) with axle ride travel.

   For a road-racing car with very stiff springs, the difference between the two arrangements should be slight, always with the proviso that "other things are kept equal".

   For the desert racer it is probably desirable to minimize the coupling between ride travel and body rolling moment. Attaching the center link to the chassis will partially fix the dimension between the sprung mass CG and the roll axis H. This also constrains the axle to move in a vertical path, avoiding lateral motion of the axle as it travels over bumps. This can be seen as fixing the rear end of the roll axis, which contributes to the dimension H in Figure 18.1.

   The second choice, fixing the center link (and thus the roll axis) to the axle, will keep the roll center at a constant height above the ground (z_{RR} in Figure 18.1). This will equalize the axle load transfer in a constant turn, regardless of ride position. On a rough road, H will change continuously, resulting in a vehicle that is continuously changing roll angle during a constant lateral acceleration turn.

   The third case, a Watt's link under the axle, behaves like the first case and allows a very low roll center (if this is desired).

   The above discussion completely ignores the contribution of the springs and anti-roll bar(s) to the vehicle roll angle, as well as the effect of front roll center movement on the CG to roll axis dimension.

17.4 Comments on Simple Measurements and Experiments

1. While there may be some camber change (negative camber as the wheel moves in the bump direction), this will usually be quite small on street cars. From a plot of camber versus ride height the approximate swing arm length for the suspension can be determined. It will probably vary at different ride heights. If the car has a strut-type front suspension, any lateral deflection of the strut towers will directly affect the camber. It may be possible to measure deflection between the strut towers (open or remove the hood) when the car is loaded (the towers will deflect toward each other).

CHAPTER 18

Wheel Loads

Solutions to this chapter's problems begin on page 213.

18.1 Problems

1. Consider a 2000 lb. vehicle with equal front and rear tracks, t, and wheelbase, ℓ.

   a. Assume all four wheels have the same static weight. Perform the calculations to show that the CG is located at mid-track and mid-wheelbase.

   b. Now assume you know the CG is located at mid-track and mid-wheelbase. Perform the calculations needed to determine the wheel loads.

2. Consider the following vehicle:

   - Total weight, W = 3700 lb.
   - Wheelbase, $\ell = 104.5$ in.
   - Front and rear tracks, t = 60 in.
   - 57/43 Front/Rear weight distribution
   - CG height of the vehicle, h = 20 in. above ground

Perform the following calculations:

a. Calculate the static normal load on all four tires. Assume x-axis symmetry.

b. Calculate the longitudinal load transfer when the vehicle accelerates at a constant 0.4g longitudinal acceleration.

c. Calculate the lateral load transfer when the vehicle accelerates at a constant 0.4g lateral acceleration.

d. Which weight transfer is greater, longitudinal or lateral?

3. Which would have more lateral load transfer, an F1 car or a double-decker bus, at the same lateral acceleration? Which would have more longitudinal load transfer at the same longitudinal acceleration?

4. Consider a 3500 lb. car traveling at 150 mph on a 600 ft. radius corner banked at 20 deg. What is the normal force on the car relative to the banking?

5. Plot the front and rear lateral load transfer sensitivity as a function of front roll center height. Use the simplified equations for lateral load transfer and body roll angle given at the top of RCVD p.683, and the terminology given in Figure 18.1 as well as in RCVD Figure 18.9, p.681. You may find it useful to develop spreadsheets or a computer program for performing these calculations.

   a. **Case 1**

      - $W = W_S = 2000$ lb.
      - $t_F = 60$ in.
      - $t_R = 60$ in.
      - $a_s/\ell = 0.50$
      - $h = 20$ in.
      - $K_F = K_R$
      - z_{RF} varies from 0 to 16 in.
      - $z_{RR} = 4$ in.

   b. **Case 2** is identical to Case 1 except that $K_F = 1.25K_R$.

   c. **Case 3** is identical to Case 1 except that $K_R = 1.25K_F$.

   d. **Case 4** is identical to Case 1 except that $a_s/\ell = 0.60$ (a longitudinal CG shift).

6. With each printing of RCVD minor changes are made to correct small errors. As of 2002 RCVD entered its fifth printing (the printing is indicated on the back of the title page). These changes are noticed only by the keenest readers.

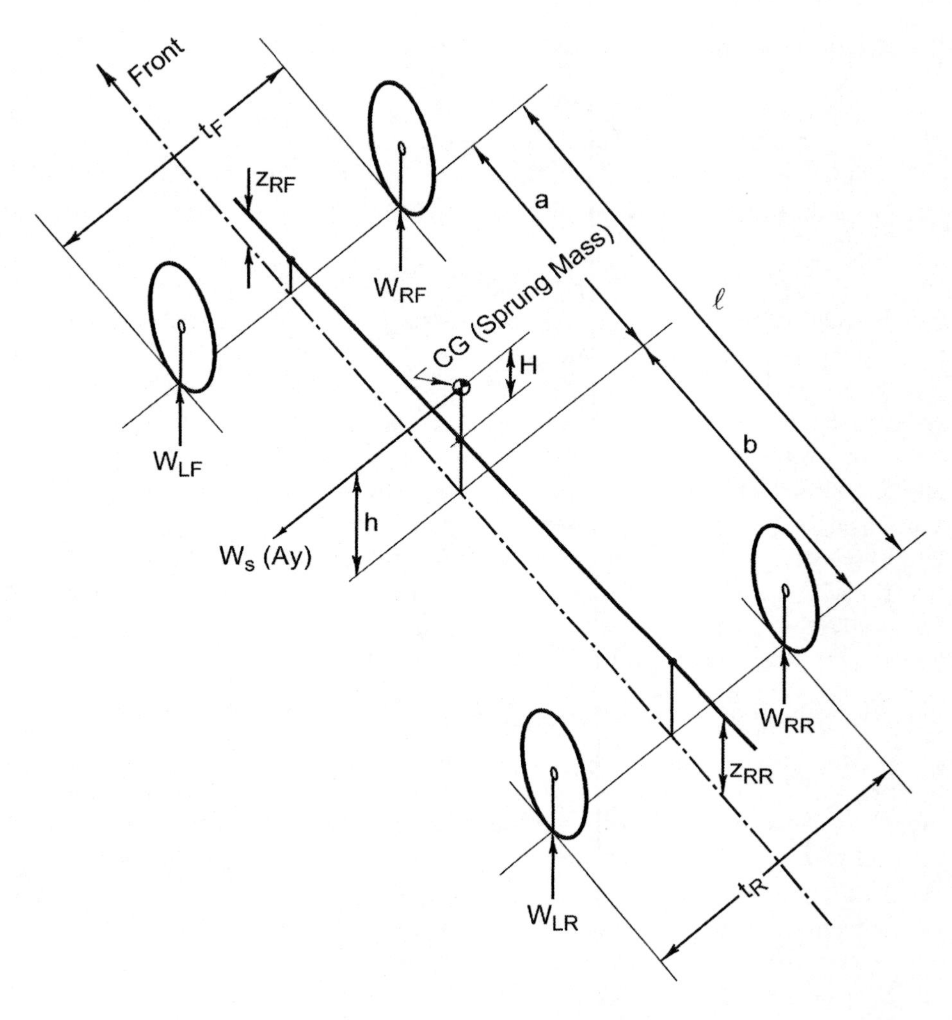

Figure 18.1 *Simplified car for load transfer calculations*

On RCVD p.682 there are two equations for the front and rear lateral load transfers, $\Delta W_F/A_Y$ and $\Delta W_R/A_Y$. The gravity term in this equation (load transfer due to lateral CG offset with roll angle) has been updated several times in an attempt to properly distribute the lateral load transfer between the front and rear axles. Derive the correct equations for $\Delta W_F/A_Y$ and $\Delta W_R/A_Y$. Compare against the equations on p.682 of your copy of RCVD. If they are different, what is the percent error of the equation in RCVD?

7. The quote at the start of RCVD Chapter 11 by Tony Rudd says, "The torsional stiffness measured ... was 240 lb.-ft./deg". Is this an acceptable figure? Why is this measure important?

18.2 Simple Measurements and Experiments

The calculations in this chapter assume a rigid chassis so that changes made at one end of the car predictably affect vehicle performance. Here is one suggested method for studying frame design and torsional stiffness, contributed by Prof. Wayne M. Brehob at Lawrence Technological University.

> The purpose of this letter is to tell you about a model construction technique that I happened on to some years ago when I had a team of ME students trying to design a frame for a Formula SAE car. We constructed model frames from soda straws by hot gluing them together. We tested these 10 inch or so long models on a small torsional test rig to obtain quantitative results and found that the model numbers "tracked" the FEA that was run on the same designs. If we made a change that improved the model, say 25%, we would see nearly the same improvement in the FEA results. One advantage was the speed with which changes could be tested—just hot glue in another tube, hang the weights on the moment arm and read the dial indicator. Other advantages were the feel for the problem that the students and I got and the fact that they could try ideas at home on the kitchen table if they had an idea, but didn't have access to the computing power.
>
> When it became clear how well soda straw models worked, I decided that a more "engineering" material like the aluminum tubing in hobby stores, would be even better. Wrong. The tubing was so stiff (high Young's modulus) that the hot glue joints broke before the deflection was large enough to see where the deflections were occurring. The low stiffness plastic in the soda straws were much better. Sometimes I have used cocktail straws if their greater length-to-diameter ratio was an advantage.

Another contributor[1] suggests balsa wood for spaceframe chassis models. Stiffness can be changed by varying the size of the members to suit the model scale and produce the required deflection.

Along the same lines, car companies use vacuum-formed plastic models to simulate stressed-panel (monocoque) structures. These models are usually larger than Prof. Brehob's, but serve the same purposes—visualization and a check on FEA modeling.

1. Design a simple ladder or spaceframe structure that could be used as an automotive chassis. Make one or more of the diagonal members (X-brace in the case of

[1] Dave Kennedy.

the ladder frame) removable. Build a simple fixture to test torsional and bending stiffness—and run tests with and without the diagonal members installed.

2. Locate four scales suitable for your vehicle (these could be bathroom scales if the vehicle is a go-kart). Calibrate the scales against a known weight or against each other. Weigh the vehicle with the scales located on a flat surface. Is the vehicle symmetric? If not, what change(s) could be made to make it symmetric?

18.3 Problem Answers

1. a. The problem states that $F_{zLF} = F_{zRF} = F_{zLR} = F_{zRR} = 500$ lb. Take moments about the front axle to determine a:

$$Wa = F_{zR}\ell \quad \Rightarrow \quad 2000a = (500+500)\,\ell \quad \Rightarrow \quad a = \ell/2$$

Now take moments about the left side wheels to determine the CG location, y, measured from the line connecting the left side wheels:

$$Wy = (F_{zRF} + F_{zRR})\,t \quad \Rightarrow \quad 2000y = (500+500)\,t \quad \Rightarrow \quad y = t/2$$

Thus, the CG is located at mid-track and at mid-wheelbase.

b. The problem states that the CG is located at $(a, y) = (\ell/2, t/2)$. There are four unknowns (the wheel loads) so four simultaneous equations will be needed. The first three are straightforward, the result of summing the wheel loads, taking moments about the front axle and taking moments about the left side wheels, respectively:

$$F_{zLF} + F_{zRF} + F_{zLR} + F_{zRR} = 2000$$

$$(F_{zLR} + F_{zRR})\,\ell = (2000)\,(\ell/2)$$

$$(F_{zRF} + F_{zRR})\,t = (2000)\,(t/2)$$

At this point the solution is indeterminate. The required fourth equation, where k is a constant:

$$(F_{zRF} + F_{zLR}) - (F_{zLF} + F_{zRR}) = k$$

This is the equation for diagonal weight, sometimes called "wedge". To obtain equal wheel loads when the CG is in the center of the car, set $k = 0$. Then solve the four simultaneous equations for the wheel loads. The result is 500 lb. on each wheel.

What if $k \neq 0$? Different wheels loads result for the same CG location depending on the value of k. In the most extreme case, setting $k = 2000$ leads to $F_{zLF} = F_{zRR} = 0$ and $F_{zLR} = F_{zRF} = 1000$. Checking this result by taking moments about the front axle and the left side wheels as in 1a confirms that the CG is located at mid-wheelbase and mid-track for this vehicle.

Changing the spring preload changes the value of k in the fourth equation, altering the static wheel loads *without changing the CG location.* Race cars fitted with onboard weight jackers give the driver a means to adjust the diagonal weight while on the circuit, thereby changing the static wheel loads and, as a result, the handling.

2. a. At rest the lateral and longitudinal accelerations are zero. Symmetry about the x-axis implies that each pair (front and rear) of wheels will be equally loaded.

$$F_{zLF} = F_{zRF} = \frac{1}{2}(0.57)(3700) = 1054.5 \text{ lb.}$$

$$F_{zLR} = F_{zRR} = \frac{1}{2}(0.43)(3700) = 795.5 \text{ lb.}$$

   b. The longitudinal load transfer is:

$$\Delta W_X = \frac{h}{\ell} W A_X = \frac{20}{104.5}(3700)(0.4) = 283.2 \text{ lb.}$$

   c. The lateral load transfer is:

$$\Delta W = \frac{h}{t} W A_Y = \frac{20}{60}(3700)(0.4) = 493.3 \text{ lb.}$$

   d. The lateral load transfer is greater, since the track is less than the wheelbase.

3. The track width of an F1 car is nearly the same as the track width of a double-decker bus. The bus, however, has a much higher CG and so it will have more lateral load transfer.

The longitudinal load transfer is not as clear. While the CG of the bus is much higher than the F1 car, the bus also has a much longer wheelbase. Since longitudinal load transfer depends on the ratio h/ℓ, a general statement cannot be made as to which has more longitudinal load transfer.

4. At 150 mph on a 600 ft. radius, the car has a lateral acceleration of:

$$A_Y = \frac{V^2}{Rg} = \frac{(150 \times (22/15))^2}{(600)(32.174)} = 2.50\text{g}$$

This gives 8775 lb. of centrifugal force. Resolving the components of the centrifugal force and the static vehicle weight normal to the roadway:

$$W_e = 3500(\cos 20 \text{ deg.}) + 8775(\sin 20 \text{ deg.}) = 6290 \text{ lb.}$$

The banking nearly doubles the wheel loads of the car at this speed. It is instructive to also calculate the lateral force required from the vehicle to corner at this speed:

$$F_y = -3500(\sin 20 \text{ deg.}) + 8775(\cos 20 \text{ deg.}) = 7050 \text{ lb.}$$

The tires must provide just over 7050 lb. of lateral force in the plane of the road (to the left, so strictly speaking this is a negative lateral force in the SAE tire axis system). The banking decreases the required lateral force by 20% over the same radius corner with no banking.

5. To begin, note that h_2, the distance from the CG to the neutral roll axis, is a function of the roll center heights and the CG location. Use the following formula:

$$h_2 \approx H = h - z_{RF} - \frac{a}{\ell}(z_{RR} - z_{RF})$$

This equation assumes that the small angle approximation applies for the inclination of the roll axis. Plots for each of the four cases are presented below.

You may find it instructive to generate plots of lateral load transfer against other parameters in the equations. For example, generate plots of $\Delta W_F/A_Y$ and $\Delta W_R/A_Y$ versus rear roll center height z_{RR}, front roll stiffness K_F or CG to roll axis distance h_2. The use of a spreadsheet or computer program simplifies the generation of such plots.

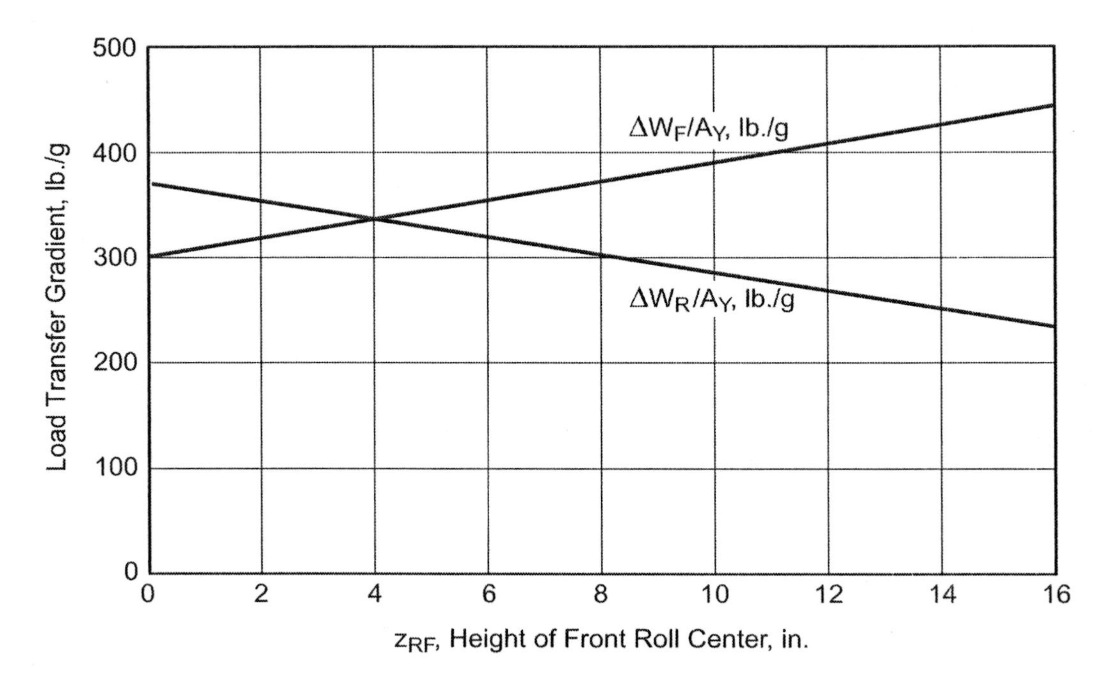

Figure 18.2 *Case 1–Equal front and rear roll stiffnesses*

a. **Case 1**—See Figure 18.2.

b. **Case 2**—See Figure 18.3.

c. **Case 3**—See Figure 18.4.

d. **Case 4**—See Figure 18.5.

6. Consider the lateral load transfer across the front axle. There are four contributors which, when summed, give the load transfer $\Delta W_F/A_Y$. The lateral acceleration of the unsprung mass:

$$\Delta W_F = \frac{z_{WF}}{t_F} W_{uF} A_Y$$

The lateral acceleration at the roll axis below the CG, proportional to the front sprung weight:

$$\Delta W_F = \left(\frac{\ell - a_s}{\ell}\right) \frac{z_{RF}}{t_F} W_s A_Y$$

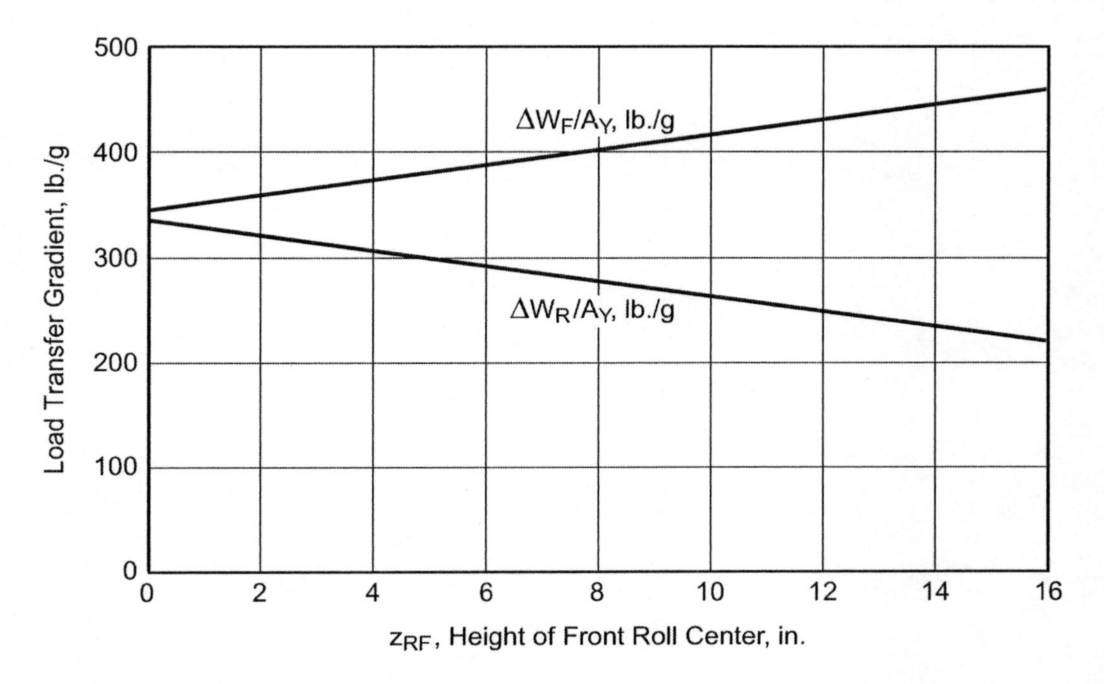

Figure 18.3 *Case 2–Front roll stiffness 25% more than on rear*

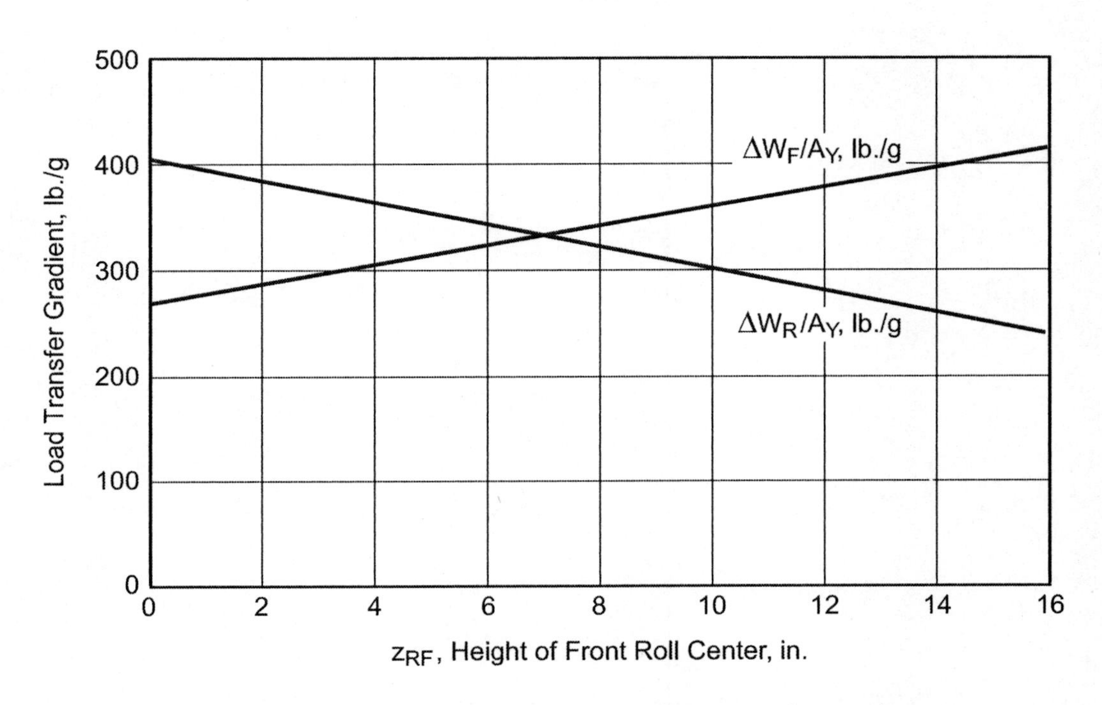

Figure 18.4 *Case 3–Rear roll stiffness 25% more than on front*

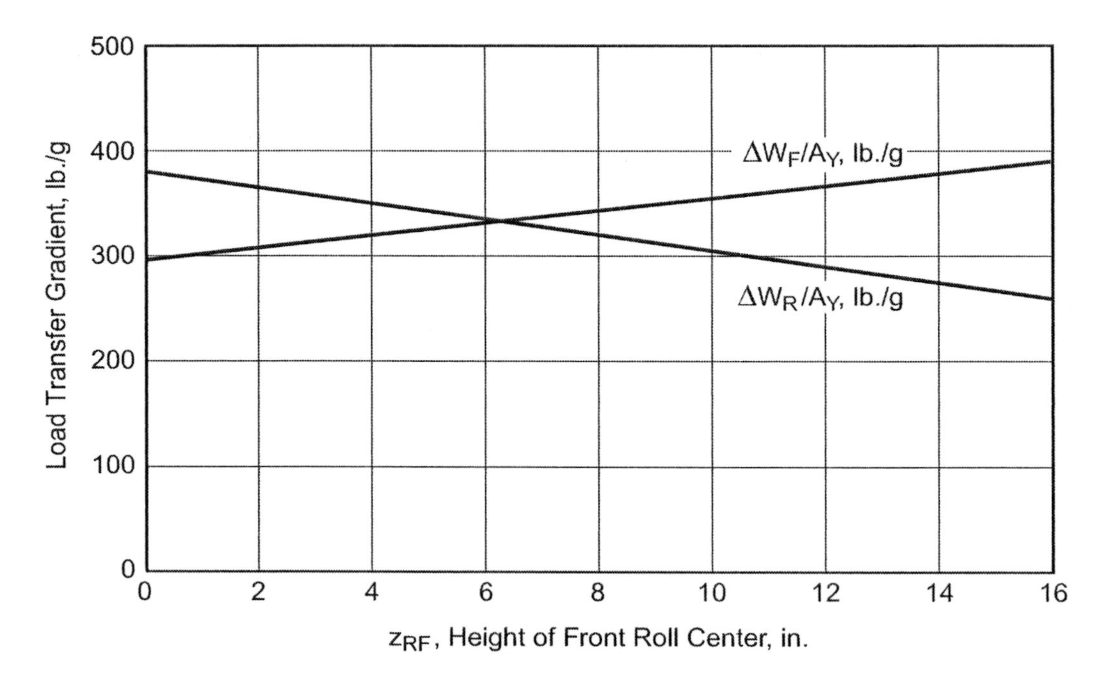

Figure 18.5 *Case 4–Equal roll stiffnesses, aft CG*

The moment about the roll axis due to lateral acceleration at the CG:

$$\Delta W_F = \left(\frac{K_F}{K_F + K_R}\right)\frac{h_2}{t_F} W_s A_Y$$

The gravity term due to lateral CG displacement with roll angle, with the roll angle expressed in radians and using the small angle assumption:

$$\Delta W_F = \left(\frac{\ell - a_s}{\ell}\right)\frac{h_2}{t_F} W_s \phi$$

But, the uppermost equation on RCVD p.682 (fourth printing) expresses the roll angle as:

$$\phi = K_\phi A_Y = \frac{-W_s h_2}{K_F + K_R - W_s h_2} A_Y$$

$$\Delta W_F = \left(\frac{\ell - a_s}{\ell}\right)\left(\frac{W_s h_2}{t_F}\right)\left(\frac{-W_s h_2}{K_F + K_R - W_s h_2}\right) A_Y$$

Summing these four contributions:

$$\Delta W_F = \left(\frac{K_F}{K_F + K_R}\right)\frac{h_2}{t_F}W_s A_Y + \left(\frac{\ell - a_s}{\ell}\right)\left(\frac{W_s h_2}{t_F}\right)\left(\frac{-W_s h_2}{K_F + K_R - W_s h_2}\right)A_Y + \left(\frac{\ell - a_s}{\ell}\right)\frac{z_{RF}}{t_F}W_s A_Y + \frac{z_{WF}}{t_F}W_{uF}A_Y$$

Simplifying:

$$\frac{\Delta W_F}{A_Y} = \frac{W_s}{t_F}\left[\left(\frac{h_2 K_F}{K_F + K_R}\right) + \left(\frac{\ell - a_s}{\ell}\right)\left(\frac{-W_s h_2^2}{K_F + K_R - W_s h_2} + z_{RF}\right)\right] + \frac{z_{WF}}{t_F}W_{uF}$$

Similarly, for the rear:

$$\frac{\Delta W_R}{A_Y} = \frac{W_s}{t_R}\left[\left(\frac{h_2 K_R}{K_F + K_R}\right) + \left(\frac{a_s}{\ell}\right)\left(\frac{-W_s h_2^2}{K_F + K_R - W_s h_2} + z_{RR}\right)\right] + \frac{z_{WR}}{t_R}W_{uR}$$

Fortunately, the gravity term is small compared to the others terms in these expressions. The (simplified) equations given in previous printings of RCVD are less than 3% off for most vehicles. The error decreases as the roll stiffness increases and roll angle decreases.

7. RCVD p.676 indicates that a typical unibody sedan has a torsional stiffness between 4000 to 10000 lb.-ft./deg. The quoted 240 lb.-ft./deg. is ridiculously low. A chassis needs to have a high torsional stiffness so that the lateral load transfer across a pair of wheels is dependent only on the roll stiffness geometry on that given axle. Said another way, changes made at one end of the vehicle should have a predictable effect on vehicle handling. A chassis with insufficient torsional stiffness may seem unresponsive to adjustments.

18.4 Comments on Simple Measurements and Experiments

1. In general, triangulated structures will give the best stiffness-to-weight ratio because they use the material in tension and compression instead of bending.

2. Most vehicles are not symmetric for two major reasons. First, the CG may be located to one side of the vehicle due to an offset engine or other large mass (e.g., the driver). Second, the car may be "wedged"—one diagonal of the four-wheeled

car supporting more load than the other diagonal. Add the weights from diagonally opposite scales to determine this. The first can be corrected (if desired) by changing the mass distribution of the vehicle (or perhaps by changing the wheel offset with spacers). The second can be changed relatively easily by changing the preload on one spring—this is commonly done by adjusting the "jack bolts" on stock cars.

CHAPTER 19

Steering Systems

Solutions to this chapter's problems begin on page 222.

19.1 Problems

1. A small sports car with a 15:1 steering ratio and parallel steer carries 1000 lb. on the front wheels. A steering wheel rim force gradient of 10 lb./g of lateral acceleration is desired. Assuming that the tire aligning torque has dropped to zero at the cornering limit, choose steering geometry (mechanical trail) and steering wheel diameter to give this value.

19.2 Simple Measurements and Experiments

WARNING: Do not get underneath a vehicle unless it is supported on jackstands and you are sure that it cannot fall or become unbalanced.

1. Locate a pair of Weaver plates (ball bearing alignment plates) to measure front wheel steer angle and a protractor to measure steering wheel angle. Repeat the

experiment shown in RCVD Figure 19.5, p.717, using your car. Plot the results for each front wheel individually and use a linear regression to determine the steering ratio over the range of ± 180 deg. of steering wheel angle.

Is the car Ackermann, reverse-Ackermann or parallel steer?

Regress the test data for turning to the right and turning to the left separately. Is a straight line a good fit to the data? Is there any hysteresis? Is there any freeplay? How can you distinguish between hysteresis and freeplay?

19.3 Problem Answers

1. Assume that the steering system is frictionless (not a bad assumption in the presence of "dither" from road roughness). Next assume a steering wheel radius of 6 in. (0.5 ft.), so the available rim force at 1.0g equates to a steering column torque of 10 lb. $\times$ 0.5 ft. $=$ 5 lb.-ft. With the steering ratio fixed at 15:1, the available kingpin torque is 75 lb.-ft.

   Since there is no pneumatic aligning torque (per the problem statement), the required trail is then given by the available kingpin torque divided by the lateral force, 1000 lb. At 1.0g lateral acceleration, 75 lb.-ft./1000 lb. $=$ 0.075 ft. or 0.9 in.

   If a larger steering wheel is used, then the trail can be increased proportionately.

19.4 Comments on Simple Measurements and Experiments

1. Most street cars will have approximately Ackermann geometry for accurate turning at low speeds and lateral acceleration. For this limited amount of front wheel steer angle, a straight regression line may give a good fit to the data. If the front wheels are steered lock-to-lock, the angularity in the steering linkage is likely to make the steering ratio nonlinear. Friction is unavoidable in steering systems so the results from left and right turning will form a hysteresis loop when the results are plotted. Any modern steering system will be preloaded internally (in the steering rack and ball joints) and have essentially no freeplay. If there is freeplay, it will appear as a change in steering wheel angle with no change in front wheel steer angle—a "flat spot" at the end of the hysteresis loop.

CHAPTER 20

Driving and Braking

Chapter 9 of *Chassis Design: Principles and Analysis*[1] contains more detailed information on driving and braking forces, including a graphical brake bias calculator, suspension design for anti-squat and anti-lift and maximum traction with several differential arrangements. Extensive calculations are included.

Solutions to this chapter's problems begin on page 225.

20.1 Problems

1. A four-wheel-drive (4WD) Miller race car was raced by Bill Milliken at Pike's Peak in 1948. At that time the road for the hill climb was loose gravel with a very low friction coefficient surface, even when dry. The Miller race car had three differentials in the drivetrain—one on the front axle, one on the rear axle plus a third center differential between the axles. This latter differential had a locking mechanism that could lock out the differential action and give a solid drive between the front and rear axles. In some respects, the Miller was a very symmetrical car—about 50% weight on each axle and equal size tires front and rear.

   During practice on the "hill" with the center differential operating, the car was very difficult to handle since wheelspin could occur on either the front or the rear, with

[1]Milliken & Milliken, *Chassis Design: Principles and Analysis*, SAE, R-206, 2002.

the car becoming very understeer or very oversteer, depending on the operating circumstances. In fact, it was probably not raceable with the center differential open. When the center differential was locked out and the average rpm of the front and rear wheels made equal, the car became quite manageable and, in fact, easier to handle than the rear or front drive cars.

This particular Miller car is now a classic machine and the opportunity to again drive it occurred but this time on the hard (but somewhat rough) surface of Bridgehampton Raceway on Long Island, NY. The surface friction coefficient at Bridgehampton must be at least twice that of Pikes Peak.

On this occasion, the car was first driven with the center differential operating. With this coefficient it was not possible to generate wheelspin in normal race-type operation. As an interesting experiment, the center differential was then locked out.

Two experienced drivers drove the car in the unlocked and locked condition. Both reported that the brake balance between the front and rear wheels (under heavy braking) was very satisfactory with the open (unlocked) center differential, the front wheels experiencing lock-up just slightly before the rears. However, in the case of the locked center differential, the car became very squirrelly under braking. The car felt sufficiently directionally unstable that neither driver would engage in hard braking at the higher speeds.

a. Explain in qualitative terms why this unstable condition under braking occurred at Bridgehampton and not at Pike's Peak when the center differential was locked out.

b. Envision a differential control system that would enable satisfactory operation on different friction coefficient surfaces.

c. The Miller weighs 2600 lb. and has a 100 in. wheelbase. Assuming a CG height of 20 in., approximate the brake bias that gave such satisfactory results at Bridgehampton? Assume a tire/road maximum μ-value of 0.85.

20.2 Simple Measurements and Experiments

> WARNING: Do not get underneath a vehicle unless it is supported on jackstands and you are sure that it cannot fall or become unbalanced.

1. Experimentally determine the type of differential used in a car.

2. Experimentally determine the final drive or axle ratio of a car and compare to the ratio given in the owner's manual. Determine the number of teeth on the final drive gears.

3. At low speed (less than 15-20 mph) run limit braking tests to determine which wheel(s) of a car lock first. Repeat on dry and wet surfaces and explain the difference.

4. If the car has ABS brakes (anti-lock), run brake tests that engage the ABS. What can you determine about the ABS control system characteristics?

20.3 Problem Answers

1. a. With the center differential operating, the car has some (currently unspecified) brake distribution between the front and rear, based upon leverage ratio and cam shape. The mechanical brakes are cable operated and the brake size is the same front and rear with single leading shoes. When braking on dry, hard surfaces, there is considerable weight transfer forward due to the relatively high CG and the high coefficient—such as at Bridgehampton.

   Even though hard braking was not a major requirement at Indianapolis (for which the car was designed), there is evidence from our experience at Bridgehampton that the brake distribution is biased toward the front. No rear lock-up occurred under quite hard braking with the center differential operating.

   When the center differential is locked out, the average front and rear wheel rotational speeds are forced to be the same. The longitudinal slip ratio, which is the ratio of the tangential speed at the bottom of the tire to the car speed, is then about the same front and rear. Since the longitudinal braking force depends on slip ratio, the ratio of longitudinal force to load on the wheel is increased on the rear and the rear tires can reach the friction limit before the fronts. In short, under heavy braking (without lock-up) the car will become loose, i.e., "oversteer-ish" or directionally unstable.

   The reason that this wasn't noticed at Pike's Peak years ago is that on dirt the coefficient is low and the longitudinal weight transfer during braking is small. On wet or icy surfaces a 50/50 brake distribution would probably be satisfactory.

   b. One possible 4WD control system enabling stable acceleration and braking on any coefficient would use a special center differential that could be momentarily

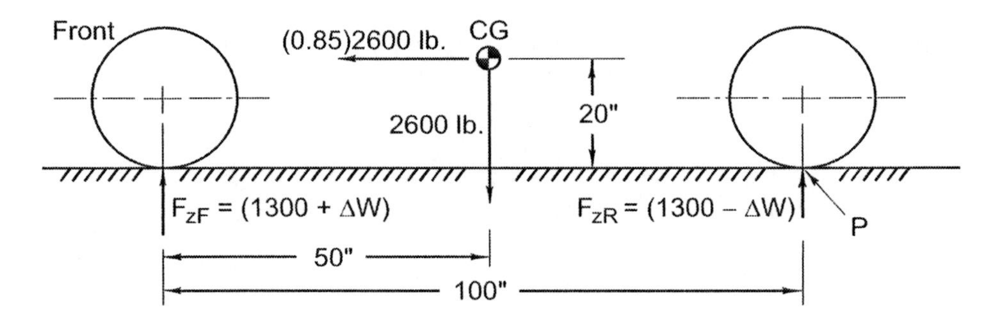

***Figure 20.1** Longitudinal load transfer during braking*

locked when any wheel started to spin-up or lock-up, based on wheel speed sensors. Hydraulic differentials which vary the input/output torque depending on sensed data could also be used to proportion the driving/braking effort to the individual tire wheel loads.

c. Take moments about point P in Figure 20.1, the rear tire ground contact point, during 0.85g braking to determine the weight transfer.

$$(\Delta W)(100) = 0.85(2600)(20)$$

$$\Delta W = 442 \text{ lb.}$$

The wheel loads are:

$$F_{zF} = 1300 + 442 = 1742 \text{ lb.}$$

$$F_{zR} = 1300 - 442 = 858 \text{ lb.}$$

Given the assumptions of this problem, the brake distribution for simultaneous lock-up front and rear is:

$$1742/2600 = 0.67 \text{ or } 67\% \text{ on front}$$

$$858/2600 = 0.33 \text{ or } 33\% \text{ on rear}$$

For a margin against rear lock-up use, say, 70% front and 30% rear.

20.4 Comments on Simple Measurements and Experiments

1. Most street cars are fitted with an open differential. With the drive wheels off the ground, lock the driveshaft or differential input by either putting the transmission in Park (automatic) or in gear (manual shift). With an open differential, when one wheel is manually rotated the other wheel will rotate an equal amount in the opposite direction. Study RCVD Figure 20.2, p.736 if this is not clear. If one wheel cannot be easily rotated then the car may be fitted with some kind of limited-slip differential. Note that this test does not load the input to the differential, so a Torsen (or a limited-slip differential with no preload) will behave like an open diff.

2. For a conventional rear drive car with a driveshaft—mark the drive tires so that the car can be pushed forward exactly one (or some other whole number) rotation of the drive tires. Place a reference mark on the driveshaft with a bright paint marker. With the transmission in neutral, slowly push the car forward and have an observer watching from the side count turns of the driveshaft. The axle ratio is commonly given by the ratio of driveshaft rotation to wheel rotation (i.e., 3.18:1). Typical pinions have around a dozen teeth, so by trial and error it is usually possible to determine the actual gear teeth from the ratio, for example 3.18:1 is likely 35:11 teeth, an actual ratio of 3.181818. . . .

   For a front drive car this measurement is much more difficult to make and may be impossible with an automatic transmission. With a manual transmission it could be done, for example, by removing the spark plugs and counting turns of the engine pulley as the car is pushed forward. This may still fail because many modern front drive transmissions do not have a direct (1:1) gear, so the ratio measured will be the product of the transmission ratio and the final drive ratio.

3. Typically one front wheel will lock before the other and both fronts will lock before either rear. This is done to prevent rear wheel lock-up, which makes a car directionally unstable. Front wheel lock-up will be more pronounced on wet or other low coefficient surfaces because the brake balance is normally set far forward to prevent rear lock-up in dry (high load transfer) conditions.

4. With all current ABS systems, once the ABS is engaged (as in a "panic stop"), the brakes will be initially "spiked" and the wheel deceleration monitored in an attempt to characterize the tire-road friction. Once the "spike" is removed, the wheels will start to roll again and will remain rolling (while the brakes are applied in an appropriate amount) until the next "spike" is applied to test the tire-road friction again. There will also be short periods when the brakes are released completely to measure the vehicle speed and compare to the deceleration that the controller has predicted from the deceleration caused by the "spike".

CHAPTER 21

Suspension Springs

Chapter 21 of *Race Car Vehicle Dynamics* contains a number of worked examples. Instead of repeating similar problems here, this chapter focuses on simple measurements and experiments.

21.1 Simple Measurements and Experiments

> CAUTION: Springs store energy when deflected and suspension springs can store a significant amount of energy. When measuring spring rate the spring must be restrained so it cannot slip out of the test fixture.

1. Locate a few springs, either torsion bar and/or coil. *Calculate* the spring rate for each. The rate may be linear or nonlinear depending on the actual spring.

2. Locate a universal (tension/compression) test machine and *measure* the spring rate for each of the springs used above. Plot the force deflection curve using 5 to 6 points for each spring.

3. Explain any discrepancies between calculation and test.

21.2 Comments on Simple Measurements and Experiments

1. Use machinist tools (micrometer, dial caliper, etc.) to measure the spring geometry. In the case of coil springs, careful inspection may be required to determine the actual number of coils that are active. The formulas to calculate spring rates from the measured dimensions are given in Chapter 21 of RCVD.

2. The test can be run in both directions (compression and extension). If multiple measurements are made over a range of deflections then a regression line can be calculated using all the data to average out any measurement errors. The data may be linear or nonlinear, depending on the spring, and the data may show a hysteresis loop which is almost certainly an artifact of the measurement process.

   There are two possible explanations for the existence of a hysteresis loop. Either the test exceeded the yield strength of the spring material (causing the spring to take a permanent set) or there is friction or another source of error in the system.

   For coil springs: During the test, note any coil binding (reduction of the number of active coils) that occurs as the test progresses.

   For torsion bars: A special fixture may be needed to constrain the torsion bar in a standard (linear motion) universal test machine.

3. Spring formulas normally assume constant wire and coil diameters. Real springs can only approach the ideal specified by formulas. For example, manufacturing variations in wire diameter cause a real spring to deviate from the theoretical rate. Some suspension springs have complex geometry and may be designed to have a nonlinear rate. Some coil springs have tapered wire or conical coil diameter and some anti-roll bars (torsion bars) are not straight. Advanced analysis (for example FEA) may be required to calculate spring rates for these springs.

   Coil springs: If coils bind during the test, the spring rate (slope of the force-deflection curve) will increase and a simple linear regression line will not give a good fit to the data, even if the spring is otherwise linear. Advanced analysis (for example FEA) may be required to calculate spring rates for these types of springs.

   Torsion bars: If the spring angular deflection is not measured at the ends of the spring, the test may measure deflections that occur elsewhere in the system.

   Both Olley[1] and the Spring Manual[2] discuss a variety of springs including leaf springs. See these references for more information.

---

[1] Milliken & Milliken, *Chassis Design: Principles and Analysis*, SAE, R-206, 2002.
[2] *SAE Spring Design Manual*, AE-11, SAE, Warrendale, PA, 1990.

CHAPTER 22

Dampers (Shock Absorbers)

Solutions to this chapter's problems begin on page 232.

22.1 Problems

1. [1]A quarter-car weighs 1000 lb., has static deflection of 6 in. and the damper leverage is 0.8. (Leverage is the ratio between the motion of the damper and the motion of the wheel.)

   a. What value of damping constant (damping force per unit velocity) in the damper would give a damping ratio of 0.25?

   b. If the wheel hit a pothole which gave it an instantaneous velocity of 30 in./sec., what force would be transmitted to the unsprung mass at that instant?

2. Describe the effects on ride and handling of damper end rubbers. Why is their nonlinearity so useful?

3. RCVD Figures 22.24 and 22.26, pp.805-6, are both generated from the same damper test. What crankshaft rotational speeds were used for each of the five test runs?

[1]This question contributed by Tony Best of Anthony Best Dynamics Ltd.

4. [2]Discuss the compromises involved in choosing a damping ratio setting for an idealized suspension, consisting of a linear spring and damper between the sprung and unsprung masses.

22.2 Simple Measurements and Experiments

1. If you have access to a damper (shock absorber) dynamometer, characterize the dampers on your street or race car.

2. Even without a damper rig, slow speed tests can be run on a universal tension-compression test machine (often found in college engineering labs) to determine the friction and low speed damping characteristics.

3. On a universal tension-compression tester, rate some damper end rubbers (relaxation springs). Is this a useful measurement?

22.3 Problem Answers

1. The question describes the geometry shown in Figure 22.1. In this figure, K is the wheel center rate, c is the damper's damping constant and c_W is the damping measured at the wheel center.

   a. Suppose the wheel center experiences a vertical velocity, v. Due to leverage, the velocity at the damper will be $0.8 \times v$. The damper force is then $c \times 0.8 \times v$. Finally, the damper force measured at the wheel is again modified by the leverage:

   $$F = 0.8 \times c \times 0.8 \times v = 0.64cv$$

   And the effective damping constant at the wheel center is:

   $$c_W = 0.64c$$

[2]This problem contributed by Tony Best of Anthony Best Dynamics Ltd.

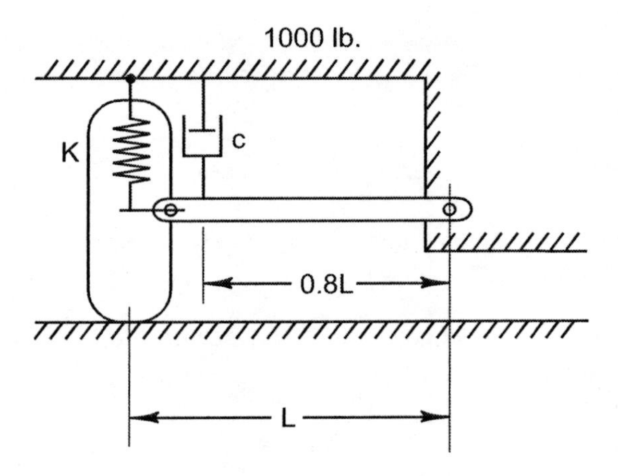

Figure 22.1 *Quarter car model showing damper leverage*

A 6 in. static deflection implies a spring rate of $1000/6 = 167$ lb./in. To achieve a damping ratio of 0.25:

$$\zeta = 0.25 = \frac{c_W}{2\sqrt{Km}}$$

$$c_W = 2(0.25)\sqrt{Km} = 2(0.25)\sqrt{\left(\frac{167}{12}\right)\left(\frac{1000}{32.2}\right)} = 10.4 \text{ lb./(in./sec.)}$$

For the actual damper, the damping constant c is $10.4/0.64 = 16.25$ lb./(in./sec.).

b. Working with the equivalent damper at the wheel center:

$$F = c_W v = 10.4 \times 30 = 312 \text{ lb.}$$

This is a force in the plane of the wheel. Forces are actually applied to the sprung mass at the actual damper end and at the effective pivot of the linkage to the body. It is the resultant of these two forces that is 312 lb. in the plane of the wheel.

2. Damper end rubbers (relaxation springs) have the following effects:

Ride—End rubbers have a minimal effect on low frequency performance (where amplitudes are large). They can improve high frequency isolation where amplitudes are low and dampers tend to be stiff.

Handling—End rubbers have little effect on primary ride motions (body natural frequency). Soft end rubbers can reduce the damping at wheelhop frequency, allowing for rapid changes in tire load and reducing road holding.

The nonlinearity of damper end rubbers allows them to be relatively soft for some initial small deflection, reducing high frequency vibration transmission to the chassis, while further deflection changes the spring rate to a very high value, forcing the damper to move (and perform its function).

Note: Do not confuse the damper end rubber with the bump stop. See RCVD Figure 22.23, p.804, for a sketch showing both elements.

3. Assume the simplest configuration: One end of the damper is connected to the crankpin and the other end is attached to ground. The peak linear velocity will be the same as the circumferential velocity of the crankpin (at the point where the crank throw is at right angles to the damper). If a "Scotch yoke" mechanism was used (crankpin running in a slider) the output would be pure harmonic motion as well.

   For ω in rad./sec., the rotational velocity of the crankpin would be $\omega = V/r$. For ω in Hertz (cycles/sec.), $\omega = V/2\pi r$. For ω in rpm, , $\omega = 60 \times V/2\pi r$.

   From the x-axis (deflection) of RCVD Figure 22.24, the radius, r, of the crankshaft (amplitude) is about 13.5 mm. Read the peak rebound forces from each loop off the same figure. Note that it is necessary to remove the 250 N offset of the gas spring to arrive at the actual damping force in Newtons. Use RCVD Figure 22.26 to determine the velocity that resulted in the peak rebound force for each loop.

Loop ID	Force N	Velocity mm/sec.	ω Hz	ω rpm
Slowest	~ 0	~ 0	~ 0	~ 0
	−700	175	2.06	124
	−1750	350	4.13	248
	−2450	520	6.13	368
Fastest	−2950	630	7.43	446

4. The spring-mass system consisting of the sprung mass on the ride springs is subject to disturbances when the unsprung mass moves in response to road irregularities. Without damping, the sprung mass response would persist indefinitely and would be very high for inputs around the natural frequency. The effect of damping is shown on Figure 22.2, on which the relative response is plotted as a function of the frequency ratio (disturbing/natural) and the damping ratio

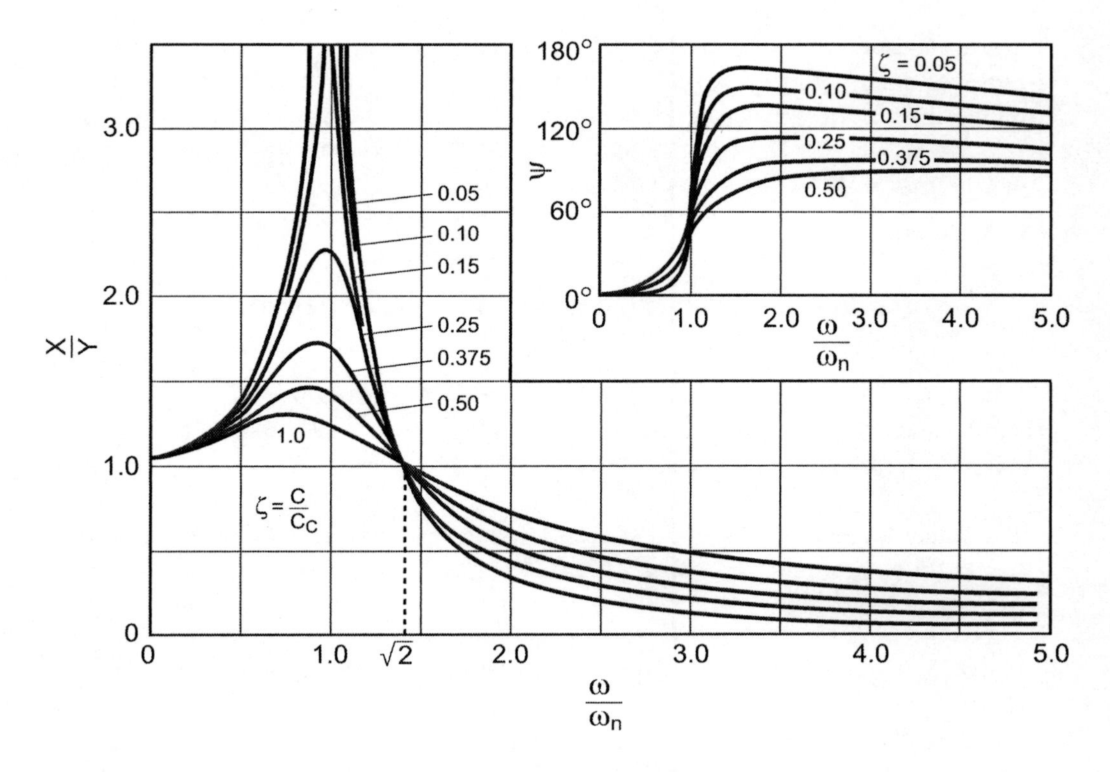

***Figure 22.2** Transmissibility*

(actual/critical). From the graph, it can be seen that the primary ride resonant response is reduced as the damping is increased. All the damping curves cross at a frequency of $\sqrt{2}$ times the natural frequency. Above this frequency increasing the damping increases the response. Thus a compromise value must be chosen for the damping.

A further influence is the nature of the vehicle. A "taut" suspension, implying high damping, will generally improve handling and driver "feel", for the sake of which a sports car driver will accept the harder ride.

The effect of damping ratio on ride and handling has been studied by taking the following measures:

- For ride: mean vertical acceleration of the body, taking into account the human susceptibility to acceleration.
- For handling: mean load variation taking into account the tire vertical spring rate and the masses (sprung and unsprung).

These measures, illustrated in RCVD Figure 22.30, p.812, were computed for a range of damping ratios for a vehicle traveling on a rough road. The optimum damping ratio for ride was approximately 0.15 while the optimum for road holding was approximately 0.45 (for that particular test case). The damping ratio was more critical for ride than handling in this example.

22.4 Comments on Simple Measurements and Experiments

1. Production or other low cost dampers are not very consistent, so testing several similar units will often give different results for each. Make sure to test over a range of velocities that you expect the damper to experience in service.

2. The force-deflection curve for a standard double tube damper at low speed should be roughly "rectangular"—constant force when moving and some breakout friction before motion begins. For gas (pressurized) shocks, the gas pressure will produce a constant force due to the differential area of the piston and rod. Depending on the gas volume there will also be a spring rate associated with damper compression.

3. Rubber in compression gives a highly nonlinear spring rate. A low speed measurement of the rate is only one of many tests required to characterize a rubber spring—other tests need to be run at different velocities and strokes as well as tests over the full temperature range. Thus, a single low velocity test is not all that useful.

CHAPTER **23**

Compliances

Solutions to this chapter's problems begin on page 238.

23.1 Problems

1. RCVD Figure 23.1, p.837, shows a machine for measuring kinematics and compliances. In this design the car chassis is fixed to "ground" and the wheel platforms are moved.

   a. Describe another arrangement for making these measurements.

   b. Compare the differences between the two machine concepts.

2. Assume that you have access to a generalized kinematics and compliance test rig which allows you to make any desired force and/or displacement measurement. Describe two different test methods for determining the front-view roll center location of an independent suspension. You may wish to refer to RCVD Sections 17.2-3, p.610-620.

3. As noted on RCVD p.834, the algebraic sign of a compliance effect is often denoted by its effect on the car—understeer or oversteer. Sketch a compliant rear suspension (with rubber isolators) that can be either US or OS (steady-state), depending on the stiffness chosen for the isolators.

4. The upper plot in RCVD Figure 23.2, p.838, was made by applying equal lateral forces to the two tires (as shown in RCVD Figure 23.1). This is called "parallel" force application. The lateral force compliance steer test is also run with forces in "opposite" directions, where one of the force vectors changes sign. Discuss the differences between parallel and opposite tests for a front suspension and steering system.

5. Sketch plots similar to RCVD Figure 23.2 that might result from a ride rate test (wheels moved vertically in bump and rebound). Make two sets of plots, one corresponding to a test where the dampers (shock absorbers) are installed and a second test where they are removed. Include the additional spring rate effect of a bump stop (jounce bumper), as shown in RCVD Figure 23.1.

6. Aligning torque compliance has a prominent effect on understeer/oversteer as defined by SAE. Why?

23.2 Simple Measurements and Experiments

1. Sketch a concept design for a simple rig that could be used to determine roll center height (and possibly lateral position). Extend the design to also measure lateral force compliance effects (steer and camber).

2. Perform some simple qualitative compliance tests on a car (or truck). Use your imagination to determine simple force inputs to load the wheels in different directions and observe the response.

23.3 Problem Answers

1. a. Attach the sprung mass of the car to a structure that can move in heave, roll and pitch. Support the wheels on multi-axis load cells, all mounted in a horizontal plane. Arrange for each wheel station to move longitudinally and laterally and to rotate (steer)—to apply tractive forces, cornering forces and aligning torques, respectively.

   b. Measured data should be similar regardless of the machine type, provided that the measurements are all referenced the same way. Small differences have to do with the motions of the unsprung masses. For example, on the machine described above, the wheels actually camber relative to the ground as on the road, so the gravity terms and tire deflections are more nearly correct.

2. Kinematic determination of the effective swing arm for each wheel is made by determining the path of the tire contact patch as the suspension is deflected vertically (in ride). Normals to the path will intersect at the instant centers. The roll center is at the intersection of the two swing arms.

   Force determination of the roll center is accomplished by applying a lateral force at the tire contact patch while noting the change in vertical load. This can be done while the suspension is deflected vertically. The resultant force (by vector addition) will pass through the roll center.

   These two measurements are often used to check each other. The "imperfections" in the suspension, such as friction and deflection of rubber bushings, are enough that the two methods never quite agree. Note that when the roll center is near the ground, the normal load changes will be small and difficult to measure reliably (theoretically, zero load change occurs if the roll center height is zero).

3. The tri-link strut rear suspension, RCVD Figure 17.29b, is amenable to this type of U/O tuning. Assume that the rubber bushings at the pivots of the forward link are softer (lower spring rate) than those in the rearward link. In this case, the lateral force from the tires steers the wheel toward the turn center (to the right in a right hand turn) and the effect is steady-state understeer.

   If the bushings are swapped front to rear (placing the softer bushings at the rear), the effect is steady-state oversteer.

4. Parallel lateral force input simulates turning with no weight transfer from inside to outside wheel—which is approximately true at low lateral accelerations. Parallel force input will apply a lateral force to the steering rack that is a function of the geometric trail. This is used to measure steering system compliances.

   Opposite force input is used in conjunction with parallel input, to simulate differences in left and right wheel lateral forces at high lateral acceleration. Opposite forces will not tend to move the steering rack, so the steer compliances measured in this case are more likely due to suspension deflections, rather than steering system deflections.

5. The curves might look like those plotted in RCVD Figure 23.2 (top), except that the X-axis is ride travel with bump to the right and the Y-axis is wheel load. The average slope is spring rate. With dampers installed, the hysteresis loop will open up. Addition of bump stops will increase the slope (spring rate) on the right hand end of the curve. If the test is extended in the rebound direction until the end of the damper (or other rebound stop) is reached, the load will drop to zero at that point.

6. SAE defines understeer as measured at the steering wheel. The introduction of aligning torque compliance steer into a "theoretically rigid" steering system will require more steering wheel angle to maintain the original path. Aligning torque compliance effects are always understeer in the linear range for the front wheels.

23.4 Comments on Simple Measurements and Experiments

1. A number of designs are possible. Perhaps the simplest is to pull laterally on the chassis at different heights above ground until a height is found where there is no vehicle roll. This will not be very precise, since there is considerable friction in the suspension (i.e., ball joints) even with the dampers removed.

   For a more comprehensive machine, it must be able to either apply a lateral force while measuring the change in vertical load, or it must have instrumentation capable of measuring wheel path while the suspension is deflected in ride.

   The first design above can be modified by setting the wheels on Weaver plates (alignment turn plates) as the lateral force is applied. The Weaver plates must be modified to resist the lateral force at the center of the contact patch, but otherwise remain free in steer. As such, the lateral force is counteracted at the ground. Careful measurement of steer and camber angles versus lateral force will allow compliances to be approximated.

2. While accurate measurements of compliances require expensive laboratory test equipment, simple qualitative tests can be easily performed to observe some effects. Compliances vary from car to car and range from very small to surprisingly large. These tests may or may not result in deflections observable with the naked eye. The experiments chosen below have produced observable results on some cars.

   Simple, visual compliance tests include:

   a. With the hand brake locked (rear wheels locked) rock the car fore and aft and observe the fore/aft axle location. Does the tire move relative to the fender? If so there is longitudinal force compliance. For front drive cars, put the car in park (or in first gear with a manual transmission) and repeat with the front wheels locked. There may be some backlash or freeplay in the transmission that will have to be moved through before the longitudinal force is applied to the suspension.

   b. Lock the steering wheel in place and put the front tires on Weaver plates (ball bearing alignment plates). Force the wheels to steer left and right by pushing alternately on the front and rear sides of the tire. Many vehicles have considerable aligning torque compliance steer which can be observed with this test.

c. Oscillate the vehicle side to side (without exciting the roll mode). This will laterally deflect the tire sidewalls. Observe the lateral location of the wheels (not the tires) to see if there is any lateral motion with lateral force and/or any lateral force compliance camber.

APPENDIX A

Program Suite

The Race Car Vehicle Dynamics Program Suite is a collection of simple vehicle dynamics programs intended to illustrate some of the concepts discussed in RCVD and acquaint the user with vehicle dynamics software. Six individual programs are included.

Development of the Program Suite began in 1996 when Dr. William Rae and Edward Kasprzak instituted a "Road Vehicle Dynamics" course in the Mechanical Engineering Department of the University at Buffalo (UB). This course was based on the Millikens' book *Race Car Vehicle Dynamics*. The inclusion of software was seen as a natural extension, similar to the software Prof. Rae had written for his "Flight Dynamics" course on aircraft. At first, only the Bicycle Model was programmed. This was followed by a program called "Night at the Races", the forerunner of the "RCVD Speedway" program included in the RCVD Program Suite. A four-wheeled model of the automobile then followed. The program has been continually and gradually developed by Edward Kasprzak since 1996. Upon assuming the role of course instructor of "Road Vehicle Dynamics" at UB in 2002, he undertook a major rewrite of the programs. This has been repackaged as the Race Car Vehicle Dynamics Program Suite.

This chapter's exercises begin on page 245.

A.1 Overview of the Programs

The User's Manual for the Program Suite can be found in Adobe Acrobat (PDF) form on the accompanying CD. It contains detailed information on how to run the programs. Installation of the Program Suite is discussed in the next section. Below is a brief overview of each of the programs included in the Program Suite.

Bicycle Model—This is the linear model of the automobile discussed at length in RCVD Chapter 5. It is the simplest vehicle model for the study of vehicle handling. Load transfer, roll stiffness distribution, aerodynamics, tire load sensitivity, etc., are ignored. This model is an excellent tool for learning the basics of vehicle stability and control. Along with a time-based integration of the equations of motion, steady-state responses and stability derivatives are calculated.

Four-Wheeler—A step up from the Bicycle Model, this model includes lateral load transfer, roll stiffness distribution, aerodynamic downforce and realistic lateral force versus slip angle tire data. While still a simple model of the automobile, the nonlinearity of the tire data and the calculation of individual wheel loads provides a noticeable increase in complexity. Time-based integration of the equations of motion produces approximately 20 output variables. This model can be used to study the effect of various parameter changes on wheel loads and the effect of the wheel loads on vehicle handling.

Moment Method—This is a linearized version of the MRA Moment Method program developed by Milliken Research Associates, Inc. It uses the Bicycle Model of the automobile to calculate automobile statics and allows the user to construct C_N-A_Y diagrams as introduced in RCVD Chapter 8.

RCVD Speedway—In this program the Bicycle Model of the automobile is used to compete in a very simple "race". The goal is to circumnavigate an object of known radius in the least amount of time. By adjusting vehicle parameters the car can be tuned for optimal performance. Time-based integration of the equations of motion produces a real-time plot of the vehicle's path on the race course. This model is an excellent teaching and learning tool for the concepts of oversteer, understeer, neutral steer, steady-state cornering, stability, etc.

RCVD Speedway 2—Similar to the RCVD Speedway, this program uses the Four-Wheeler instead of the Bicycle Model. The additional complexity of the Four-Wheeler vehicle model makes vehicle optimization much more challenging. The user must deal with some of the most significant aspects of setting up a real vehicle such as load transfer, aerodynamics and nonlinear tire characteristics. This model illustrates the race engineer's task in setting-up a car, albeit on a very simple race course.

Ride and Roll Rate Calculator—This model is based on the material of RCVD Chapter 16. A single track of the automobile is presented. The user can calculate overall ride and roll stiffnesses from individual tire, spring and anti-roll bar rates, or use desired totals to specify individual components. The model is both instructive and a handy calculator for anyone doing ride and roll rate calculations.

A.2 Exercises

The exercises presented below are a *starting point* for the reader's own exploration. There are many more uses and interesting exercises for each of the programs. These exercises are intended to illustrate some program applications, as well as their limitations. Exercises are grouped according to the program.

1. ***Bicycle Model:*** Run the Bicycle Model with the default inputs. Plot the results with the World X-Coordinate on the y-axis and the World Y-Coordinate on the x-axis. This will place the vehicle heading toward the top of the screen at the start of the run. Adjust the scaling so that the graph has the same scale in each direction. Is the resulting plot a circle? Look at the tabular output to determine how long the vehicle took to reach steady-state.

   What happens if you plot World X-Coordinate on the x-axis and World Y-Coordinate on the y-axis?

2. ***Bicycle Model:*** At the start of a run, with the default initial conditions, the vehicle is traveling straight ahead. With a fixed, non-zero steer angle, a step steer input results at time zero. Use the default inputs and a fixed steering angle of 5 deg. Determine the vehicle's time constant from the graph of yaw rate vs. time.

   Why does the vehicle sideslip initially go the opposite direction of its steady-state value while yaw rate does not?

   Hint: To get a better look at the start of the run, reduce the Output Record Increment and then reduce the Length of Calculation.

3. ***Bicycle Model:*** Choosing an acceptable Calculation Increment for a simulation based on numerical integration is a compromise between solution accuracy and run time. The default value in the Bicycle Model is 0.001 sec. per integration time step. Set the Calculation Increment to the following values and run the simulation: 0.00001, 0.0001, 0.001, 0.01, 0.1, 0.5, 1 sec. For each run, set the Output Record Increment to the same value as the Calculation Increment so that the results of every time step are recorded.

Use the following vehicle: m = 60, Iz = 600, a = 4, b = 6, Standard front and rear tires. Use an initial speed of 88 ft./sec. and a fixed steer angle of 7 deg. Set the Calculation Length to 4 sec. Compare the calculated final position from each run. Also, subjectively compare run times. What is an acceptable value for the Calculation Increment?

Repeat this experiment with sinusoidal steering. Use an amplitude of 7 deg. and a period of 0.2 sec. Compare the ground paths (plot World X- and World Y-coordinates) and final positions of each run. Has the acceptable Calculation Increment changed? Why or why not?

4. ***Bicycle Model:*** Start with the default vehicle. Set the fixed steer angle to 0 deg. and input a road tilt of 20 deg. What response do you expect? Is this car neutral steer? How can you tell?

   Now move the CG to $(a,b) = (4,6)$. This is a forward CG location. Look at the path response and compare with RCVD Figure 5.22, p.169. Is this an oversteer or understeer response? Reverse a and b, $(a,b) = (6,4)$. Now what response do you see? Compare your responses with RCVD Figure 5.14, p.143.

   Suppose you wanted to produce an understeer response without moving the CG. What other changes to the Bicycle Model vehicle could you make? Test your theories with simulation runs.

5. ***Bicycle Model:*** An understeer car spirals outward at fixed steer angle with increasing speed. Show this using the Bicycle Model program. Start with the default values and set a = 4, b = 6. Use an initial speed of 5 ft./sec. and choose the "Modest Acceleration" option. Run the simulation for 50 sec. and then plot the ground path. Does it spiral outward? Also, plot the front and rear slip angles against time and the front and rear tire forces against time.

   Repeat for an oversteer car, which should spiral inward. Use a = 5.1 and b = 4.9. Is the result as anticipated? Explain the behavior shown in a plot of the ground path.

6. ***Four-Wheeler Model:*** Start with the Sedan default settings at a fixed steering angle of 6 deg. and a speed of 100 ft./sec. Conduct a run and look at the results. Now replace the left front tire with the Formula One Front Tire. Predict what will happen if you repeat the same run and then use the simulation to test your theory. Return the left front tire to the Sedan Tire and place the Formula One Front Tire on the right front. Again, predict the results of the simulation before conducting the run. Why is there such a difference between placing the Formula One tire on the left or right front?

Repeat this experiment by placing the Formula One tire on the left or right rear. Are the results as you expected? Why do they differ?

7. ***Four-Wheeler Model:*** For each of the four default vehicles, perform a test to determine their understeer gradient. Make adjustments to the inputs so that the race cars are neutral steer and the sedan has an understeer gradient of 4.0 deg./g.

8. ***Four-Wheeler Model:*** The first chapter of RCVD begins with a quote from Peter Wright: "Driving a car as fast as possible...is all about maintaining the highest possible acceleration level in the appropriate direction". In this exercise, adjust the parameters of the Four-Wheeler model to achieve the highest possible lateral acceleration in steady-state cornering. Begin with the F1 Race Car and observe the following rules:

    - The mass and moment of inertia may not be changed.
    - The wheelbase must remain at 9.67 ft.
    - The CG height and wheel track may not be changed.
    - The sum of the front and rear lift coefficients may not exceed −5 (−5.1 would be considered "illegal").
    - The frontal area may not be changed.
    - The tires may not be changed.

    Click on the F1 Race Car button of the Define Vehicle screen (to set the baseline values) and then click on the Custom button. This allows you to change the values. Run the simulation at 150 ft./sec. at a Calculation Increment of 0.01 sec. For each run, view the steady-state lateral acceleration value in the Tabular Output. Adjust the vehicle (within the rules) to maximize this value. Note that you will need to adjust the magnitude of the fixed steer angle for each vehicle configuration.

    Repeat this experiment at different speeds or with different cars.

    Note that the vehicle path is not a concern in this exercise. A more advanced problem is to adjust the vehicle parameters to traverse a given path in the shortest time. See Exercise 12 for a simplified version of this racing problem.

9. ***Moment Method:*** Look at RCVD Figure 8.11, p.318. Recreate these C_N-A_Y plots using the Moment Method program.

10. ***Moment Method:*** Select a vehicle configuration and determine the maximum steady-state lateral acceleration from the C_N-A_Y diagram. Run the same vehicle configuration on the Bicycle Model program to confirm this result.

11. ***RCVD Speedway:*** Use a trial and error procedure to determine the set-up which produces a minimum lap time around a 50 ft. radius circle. Observe the following restrictions:

    - The vehicle mass = 60 slugs, inertia = 600 slug-ft^2 and wheelbase = 10 ft.
    - The cornering stiffness at each end of the vehicle must not exceed −500 lb./deg.
    - The breakaway slip angle, front and rear, must lie between 0 and 6 deg.

    Specify the CG location (a and b), the cornering stiffnesses, the breakaway slip angles, the starting position, the velocity and the steer angle.

    Note to Instructors: This has been used as an in-class exercise for a number of years in the "Road Vehicle Dynamics" course at the University at Buffalo. This exercise is performed after the Bicycle Model and the concept of under/oversteer has been introduced, but before the closed form solution to this problem has been presented (see Chapter 5 Problem 5). The class is divided into five or six teams who are then given a few minutes to agree on a design. Designs are submitted simultaneously on a sheet of paper to the instructor and then the RCVD Speedway is run on the in-class computer. The class gets to watch the results projected on the screen at the front of the room. A "spreadsheet" is kept on the front board showing the designs and the resulting lap time. In an 80 minute class there is usually time for four heats (each with one design per team)—about 5 minutes are given between heats for the teams to determine a new design. The exercise is popular with the students and in the end the best result is typically not far from the optimum.

    Try the theoretical best value from Chapter 5 Problem 5. Why does the vehicle crash into the inside retaining wall? What adjustment could you make to prevent it from crashing? How much time do you lose from the theoretical optimum due to this adjustment?

12. ***RCVD Speedway 2:*** Using the Formula One Race Car as a starting point, determine the minimum lap time around an inner radius of 200 ft. Observe the following rules:

    - You are allowed to change the following values: Front and rear distances to the CG, front and rear downforce, front roll stiffness distribution.
    - You are NOT allowed to change any other parameters, including tires, CG height, wheelbase or wheel track.
    - The sum of the front and rear aero coefficients must be kept at −5.00.

    How easy do you find it to reduce the lap time below 10 sec.? Below 9 sec.? Can you get below 8.5 sec.?

13. ***RCVD Speedway 2:*** *The RCVD Shootout.*[1] A race is to be held on two constant radius circles. The total time to complete the race will be the sum of the individual times for a lap around each radius. While the initial speed, starting point and steer angle can (and should) be different for each radius, the same vehicle set-up must be used on each radius.

    The vehicle rules are as follows:

    - Start with the NASCAR stock car. You may NOT change the tires, mass, inertia, CG height, track width or frontal area.
    - The wheelbase must be kept at 9.16 ft.
    - The front and rear aero coefficients must sum to no more than -1.0 (that is, summing to -1.01 would be considered illegal)
    - You are allowed to change the following values: Front and rear distances to the CG, front & rear downforce and roll stiffness distribution.

    The radii are as follows:

    - Radius 1: 400 ft.
    - Radius 2: 50 ft.

    To determine the time to complete a race, settle on a vehicle design and run the RCVD Speedway 2 program on the 400 ft. radius. Adjust the driver inputs (starting point, initial speed and steer angle) until you have achieved a representative lap time. Record this lap time. Then, run the same vehicle design on the 50 ft. radius circle. Again, adjust the driver inputs until you have a lap time which shows the potential of the design. Note this lap time and add it to the lap time on the 400 ft. radius. This is the time for the race—it is the quantity you want to minimize.

    In the spirit of competition, you may submit your design and driver inputs to the MRA website: www.millikenresearch.com\rcvdpae.html. The top speeds will be posted. There are no prizes for the fastest time, only bragging rights. In the future, additional races/competitions may be posted on the MRA website.

[1] Kasprzak, E., K. Lewis and D. Milliken, "Steady-State Vehicle Optimization Using Pareto-Minimum Analysis", Paper 983083, 1998 SAE Motorsports Engineering Conference & Exposition, Dearborn, MI, November 1998.

14. ***Ride and Roll Rate Calculator:*** A certain car has the following measurements:

    - Front track = 60 in.
    - Rear track = 60 in.
    - Independent Front Suspension
    - Axle Rear Suspension with 50 in. spring track
    - Weight on the front axle = 1600 lb.
    - Weight on the rear axle = 1000 lb.
    - Tire spring rate (for each tire) = 1500 lb./in.
    - CG height = 18 in.
    - Roll axis is on the ground (no jacking force)

    Suppose ride frequencies of 1.0 Hz on the front and 1.1 Hz on the rear are desired. A roll gradient of 3.0 deg./g is also desired, with 75% of the roll stiffness on the front. Specify the necessary front and rear wheel center rates and anti-roll bar stiffnesses. Ignore the unsprung mass of the vehicle and any damping effects.

A.3 Problem Answers

1. The plot is shown in Figure A.1. To scale the axis such that the resulting plot is a circle, use the "Make X Square" button on the Manual Scaling menu.

   The plot is a circle except for the initial transient. The tabular output shows that lateral velocity and yaw rate reach their steady-state values at approximately 0.20 sec.

   Recall that the SAE vehicle axis system places the x-axis positive forward and the y-axis positive to the right. That is, the y-axis points 90 deg. clockwise from the x-axis. Plotting with the X-Coordinate along the x-axis and the Y-Coordinate along the y-axis places the y-axis 90 deg. counterclockwise from the x-axis. As a result, a right hand turn would appear as a left hand turn in the plot. To see this behavior, repeat this run with a simulation time of only 5 sec.

2. Yaw rate reaches 98% of its steady-state at 0.175 sec. This is equal to four time constants, so the time constant for the vehicle is 0.044 sec.

   When the vehicle is above the tangent speed, the lateral force from the step steer initially causes the whole vehicle to want to translate to the right. The rear tires

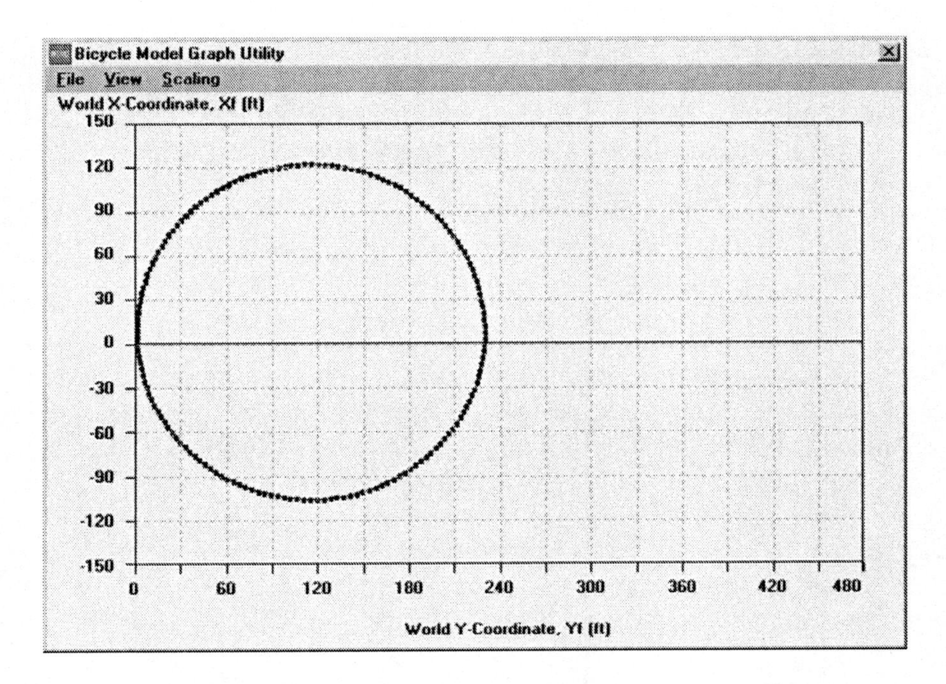

Figure A.1 *Default output with square axis scaling*

resist this motion and cause rotation of the vehicle to begin. Initially the vehicle sideslip angle, lateral velocity and the rear slip angle move opposite from their steady-state value. Above the tangent speed the steady-state attitude is "nose-in cornering", but the initial pulse can be thought of as "nose-out cornering".

Repeat this experiment at a speed below the tangent speed (say, 20 ft./sec.). Note that the vehicle sideslip angle does not reverse itself since the vehicle is always cornering with a nose-out attitude.

3. With a fixed steer angle, there is little sensitivity to calculation increment. This is because the vehicle is cornering at steady-state for most of the run. The final position after 4 sec. differs by only 5 ft. between step sizes of 0.001 sec. and 0.05 sec. Step sizes smaller than 0.001 sec. produce negligibly small increases in accuracy with an unacceptably large increase in solution time.

   The step size needs to be matched to the vehicle's natural frequency and the frequency of the inputs. Numerical integration routines used to calculate the solutions to differential equations in simulations typically assume the state derivatives are constant throughout a timestep. Higher frequencies imply that the state derivatives change value more quickly and thus require smaller calculation increments. One rule-of-thumb is to have at least 10 points per cycle of a state variable.

Step size also affects how the input signal is described. Sinusoidal steering at 5 Hz (period of 0.2 sec.) results in a reasonable Bicycle Model solution at calculation increments of 0.001 sec. and 0.01 sec., but gives no result at 0.1 sec. At 0.1 sec., there are only three points per steering cycle and, in this case, they happen to occur exactly when the steering input is zero.

Try the following: Adjust the period of the steer input to 0.205 or 0.1005 sec. and note the plot of steer angle versus time. This illustrates the "beating" phenomenon which can occur when the sampling rate of a measured signal is too low. Another effect of insufficient sampling rate can be seen with a steering input period of 0.051 sec. The resulting plot of steer angle versus time appears as a sine wave with a much larger period—this is an example of "aliasing". In signal measurement, beating and aliasing can be avoided by sampling at a sufficiently high rate.

4. The default vehicle is neutral steer. The path is a straight line at an angle to the road centerline and the vehicle sideslip is always zero. Both the front and rear slip angles grow at the same rate. The value of N_β on the stability derivatives page is zero, confirming that the vehicle is neutral steer.

   With the CG at $(a, b) = (4, 6)$ the vehicle is understeer. Figure A.2 shows the vehicle path.

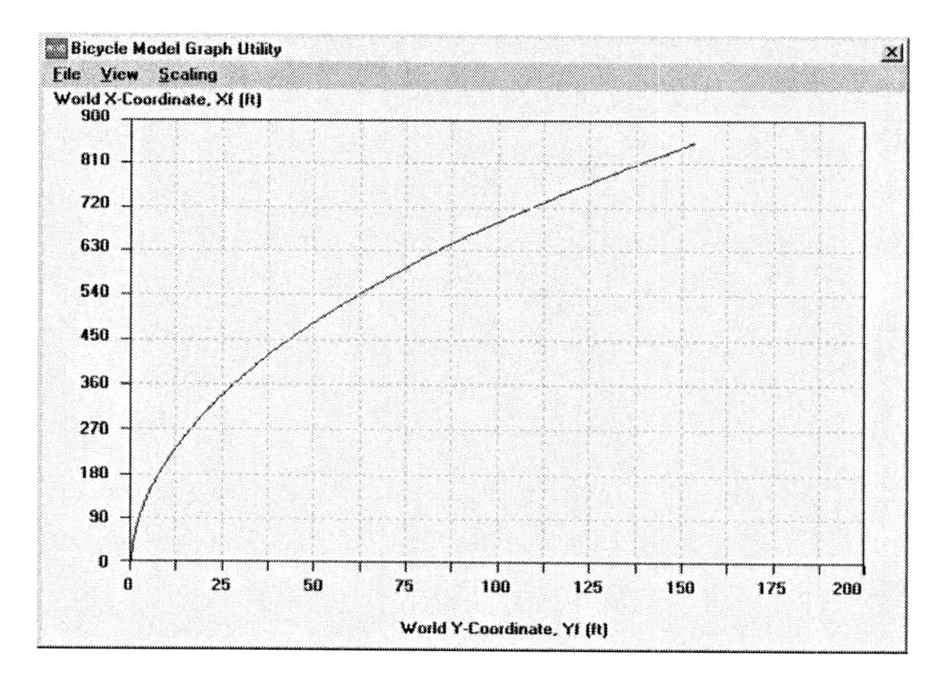

Figure A.2 *Understeer car on a tilted road*

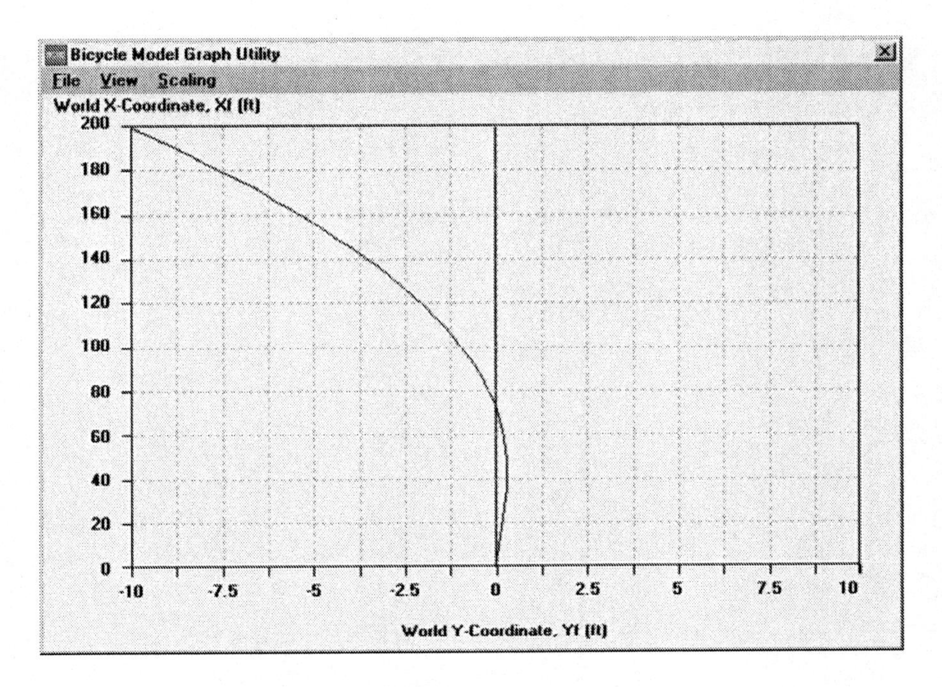

Figure A.3 *Oversteer car on a tilted road*

With the CG at $(a, b) = (6, 4)$ the vehicle is oversteer. Figure A.3 shows the vehicle path. The vehicle begins to move down the slope, but since the vehicle "oversteers" the front wheels, it turns up the slope of the road.

The tire cornering stiffnesses can be adjusted to produce understeer or oversteer. If the CG is at mid-wheelbase, placing a tire with a lower cornering stiffness on the front (compared to the rear) will result in an understeer vehicle.

5. The understeer vehicle spirals outward as shown in Figure A.4. The rear slip angle does not grow beyond 4 deg. since this angle provides the lateral force needed to keep the vehicle in yaw balance when the front tires are saturated.

   The oversteer car spirals inward until the front and rear tires are saturated. See Figure A.5. The oversteer car is unstable with fixed steering and, once the front and rear tires are over the breakaway slip angle, the vehicle will not recover. Beyond this point the simulation is not sophisticated enough to give realistic results. The simulation references speeds relative to the path—speed along the path increases at 0.1g in constant acceleration mode, regardless of the vehicle attitude. In reality, the vehicle would have spun to a stop as sketched by the dashed line in Figure A.5.

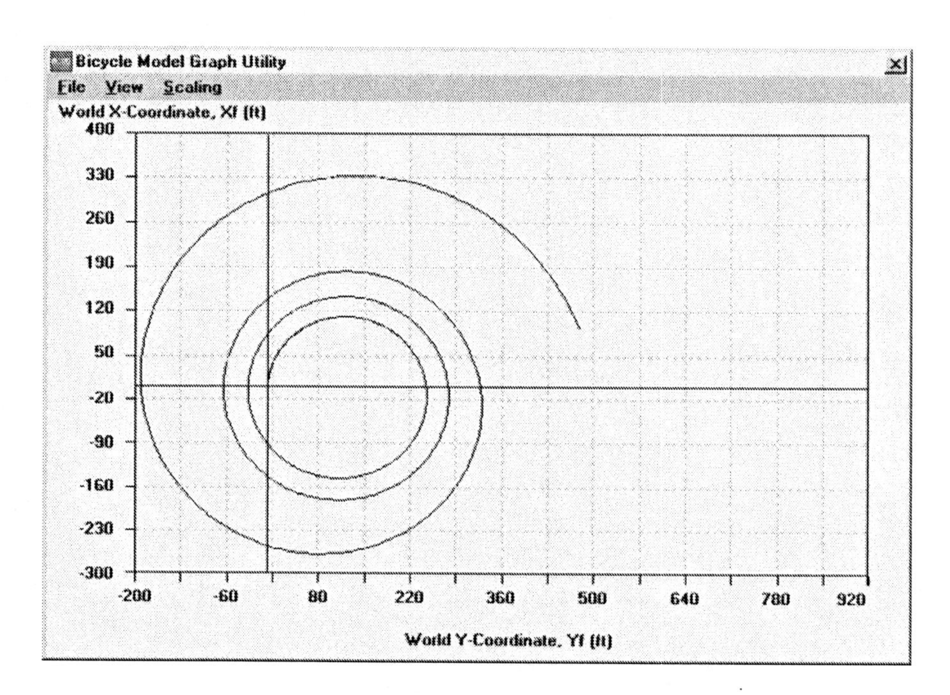

Figure A.4 *Understeer car spiraling outward*

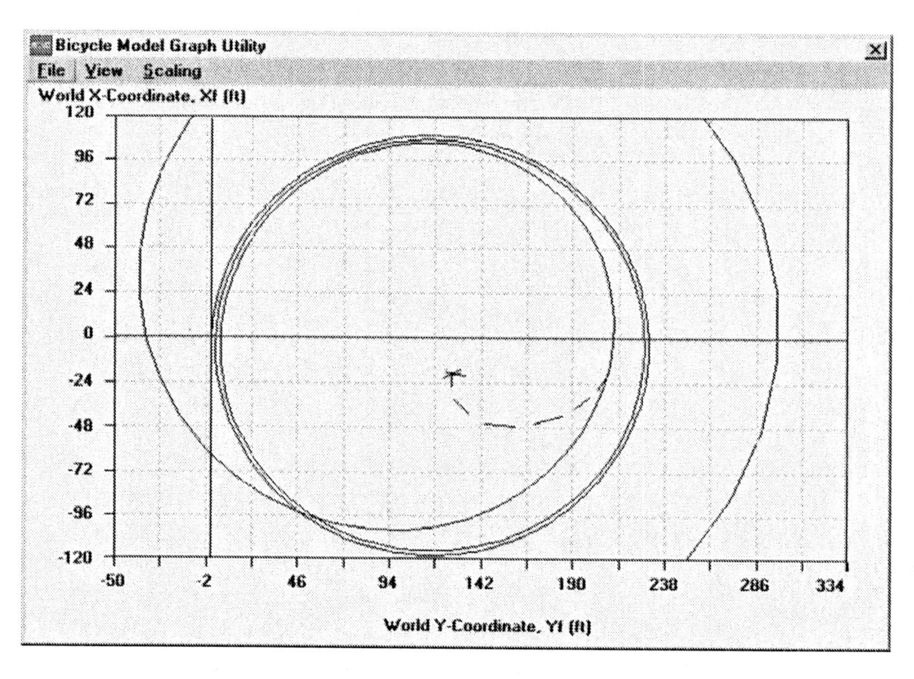

Figure A.5 *Oversteer car spiraling inward*

6. Compare the plots of yaw rate, lateral acceleration and vehicle sideslip against time for each of the three runs. For the standard Sedan, the steady-state values of these quantities are 17.16, 29.96 and −3.12, respectively.

   Placing the Formula One Front Tire on the left front causes the vehicle to spin out. This tire is much more capable than the original Sedan Tire and therefore it makes the car oversteer.

   Replacing the right front tire with the Formula One Front Tire has a less dramatic result. The vehicle does not spin out. The steady-state values are higher than for the original sedan: 18.23, 31.81 and −3.82, respectively.

   The difference lies in the amount of load carried by each of the tires. For the standard sedan in this maneuver, the outside (left) front wheel carries approximately 1500 lb. while the inside (right) wheel carries only about 200 lb. Placing a more capable tire on the inside wheel increases performance, but since the wheel load is small the gain is likewise small. A stronger tire on the outside front wheel results in a very large increase in cornering capability at the front of the vehicle—to the point that the vehicle becomes unstable.

   Placing a stronger tire on either rear wheel will have a stabilizing effect on the vehicle. Once again, the effectiveness is determined by the amount of load carried by the wheel. Since the majority of the roll stiffness of the sedan is taken across the front axle, the rear wheel loads remain very similar during cornering. As a result, the difference between placing a stronger tire on the left or right rear is not as drastic as on the front axle.

7. For each vehicle, a constant speed variable radius test can be conducted using a very slow ramp steer input. A plot similar to the bottom of RCVD Figure 5.51, p.213, can be produced. The plot for the Sedan at 88 ft./ and a ramp steer of 0.2 deg./sec. is shown in Figure A.6.

   Note that the "kinks" in Figure A.6 are due to linear interpolation of the tabular tire data. Compare Figure A.6 with the plot of the Sedan Tire. These slope discontinuities can be avoided by using a mathematically based tire model (instead of table lookup) or by using a higher order interpolation routine.

   Calculation of the understeer gradient from this plot is made using the following formula:

$$\mathrm{UG} = \frac{d\delta}{dA_Y} - \frac{d\delta_{Ack}}{dA_Y} = \frac{d\delta}{dA_Y} - \frac{\ell g}{V^2}$$

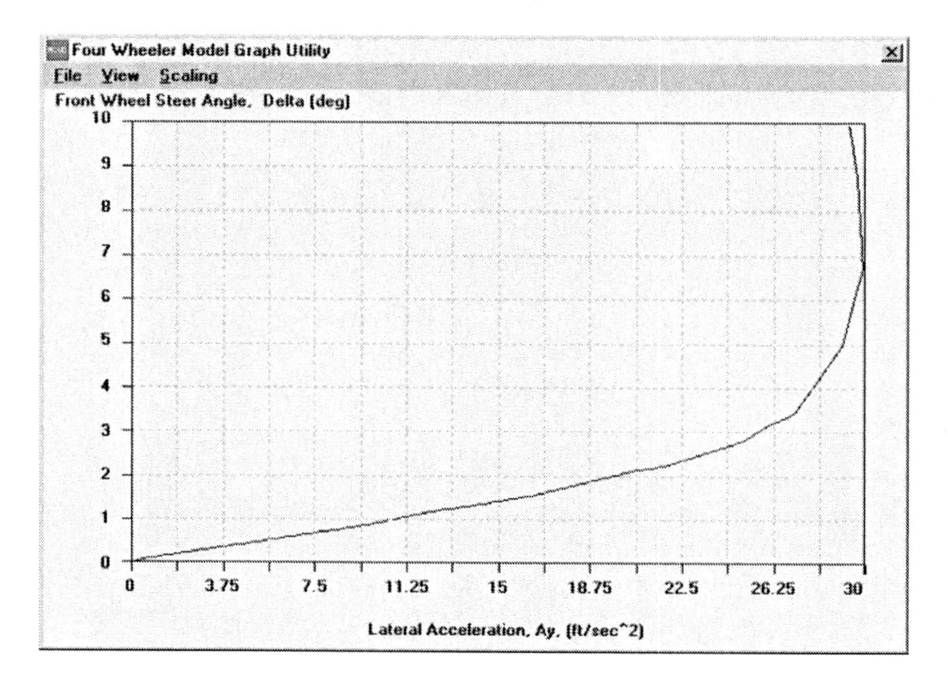

Figure A.6 *Constant speed test for the sedan to determine the understeer gradient*

For the Sedan, the understeer gradient is approximately 1.0 deg./g. A number of adjustments can be made to bring this vehicle to 4.0 deg./g. Use RCVD Chapters 5 and 12 as a guide. The same test can be repeated to determine the understeer gradient of the other three vehicles.

8. Since straight line speed is not a concern, the maximum downforce should be used. The other parameters should be adjusted so that the vehicle is neutral steer. Be careful not to lift wheels (a wheel load with zero value)—this simple model only approximates vehicle behavior when a single wheel is off the ground.

   It is useful to conduct ramp steer tests to assess your set-up changes. A modest ramp steer (perhaps 0.2 deg./sec. at the roadwheel) approximates steady-state cornering. This will eliminate the need to determine the correct steer angle—it will be apparent from a graph of steer angle versus lateral acceleration.

   It is also helpful to compare tire normal loads and slip angles with the tire data. Visualize where each tire is operating on the curve and try to move the operating points toward the peaks. You won't be able to get all the tires at the peak of the curve at the same time due to lateral load transfer. See RCVD Chapters 7 and 18. Within the rules of this exercise, a steady-state lateral acceleration over 85 ft./sec.^{2} is possible.

9. All vehicles have a mass of 60 slugs. Solutions are given for each of a-f in RCVD Figure 8.11, p.318. These are not the only possible answers—they represent the relationship among the parameters as specified in the figure.

    a. a = 5, b = 5, Standard front tire, Standard rear tire

    b. a = 5, b = 5, Standard front tire, Weak rear tire

    c. a = 5, b = 5, Weak front tire, Standard rear tire

    d. a = 5, b = 5, Standard front tire, Weak rear tire

    e. a = 4, b = 6, Race front tire, Standard rear tire

    f. a = 4, b = 6, Standard front tire, Standard rear tire

10. Consider a 60 slug vehicle with a = 4, b = 6 and Race tires on the front and rear. The Bicycle Model gives a maximum lateral accelerationof 103.4 ft./sec.^{2}. This value can be determined from the Moment Method Program by locating the point where the diagram boundary crosses the x-axis (the trim line). This point is at 3.26g, or 104.9 ft./sec.^{2}, which agrees closely with the result of the Bicycle Model.

11. The theoretical best solution occurs when the CG is mid-wheelbase, equal tires of maximum cornering stiffness are used on the front and rear, the steer angle is the Ackermann angle for the given radius and the velocity is determined from the available tire forces and the given radius. This solution will crash because there is an initial transient—the yaw rate and lateral velocity are zero at the start of the first lap. As a result, the steady-state path is a circle which is not concentric with the inner radius and the theoretical best design hits the inside wall. (Instructors may or may not want to tell their students about the transient beforehand.) To allow for this, the steer angle can be reduced slightly or the velocity increased slightly.

12. A closed form solution to this problem does not exist since the tire data is both non-linear and tabular. The problem mimics the real-world task of a race engineer to set-up a race car for a given circuit. While there are many outputs (indeed some, like slip angle, are not easily measured on a real vehicle), the information is still limited. The Bicycle Model's "driver" does not provide any feedback in this case, making the task more difficult.

    Lap times of 8.8 sec. are possible. Can you go any faster?

13. This exercise is another step up in complexity. The same vehicle design will not be optimal for both radii, due primarily to aerodynamics. Aerodynamic forces are speed dependent and will be greater on the larger radii circle.

In this exercise, the race is comprised of a single lap on each radius. If the race was comprised of m laps on one radius and n on the other, the optimum design would change once again (tending toward the optimum design for the radius with more laps on it). The disparity in optimum designs for each radius increases as the aerodynamic downforce increases. Repeat this experiment with the Formula One Car to illustrate the differences.

On a road racing circuit there may be a combination of low, medium and high speed corners. Obtaining the fastest lap time is a compromise among the cornering speeds on each size corner. A well-designed car will have little variation in optimal CG location, roll stiffness distribution and downforce distribution across the entire speed range. Access to high quality tire data is essential to create such a design.

14. Begin by determining the total roll stiffness needed. At 1g, a moment equal to the weight of the vehicle times the height of the CG should produce a 3.0 deg. roll angle. Thus:

$$K_\phi = \frac{M_\phi/A_Y}{RG} = \frac{(2600\text{ lb.})(1.5\text{ ft.})/1.0\text{ g}}{3.0\text{ deg./g}} = 1300\text{ lb.-ft./deg.}$$

The front roll stiffness needs to be 75% of this value, or 975 lb.-ft./deg., leaving 325 lb.-ft./deg. at the rear. For a 1.0 Hz front ride frequency, the front ride rate needs to be:

$$\omega_F = \sqrt{K_F/m_F}$$

$$\left(\frac{1.0\text{ cycle}}{1\text{ sec.}}\right)\left(\frac{2\pi\text{ rad.}}{1\text{ cycle}}\right) = \sqrt{\frac{(K_F\text{ lb./ft.})\left(1\text{ ft.}/12\text{ in.}\right)}{1600\text{ lb.}\Big/32.2\text{ ft./sec.}^2}} \Rightarrow K_F = 163.6\text{ lb./in.}$$

Values are then input into the Ride/Roll Rate Calculator as shown in Figure A.7. A wheel center rate of 183.6 lb./in. and an anti-roll bar providing 816.3 lb.-ft./deg. are needed on the front.

At the rear, the ride rate should be:

$$\omega_R = \sqrt{K_R/m_R}$$

$$\left(\frac{1.1\text{ cycle}}{1\text{ sec.}}\right)\left(\frac{2\pi\text{ rad.}}{1\text{ cycle}}\right) = \sqrt{\frac{(K_R\text{ lb./ft.})\left(1\text{ ft.}/12\text{ in.}\right)}{1000\text{ lb.}\Big/32.2\text{ ft./sec.}^2}} \Rightarrow K_R = 123.7\text{ lb./in.}$$

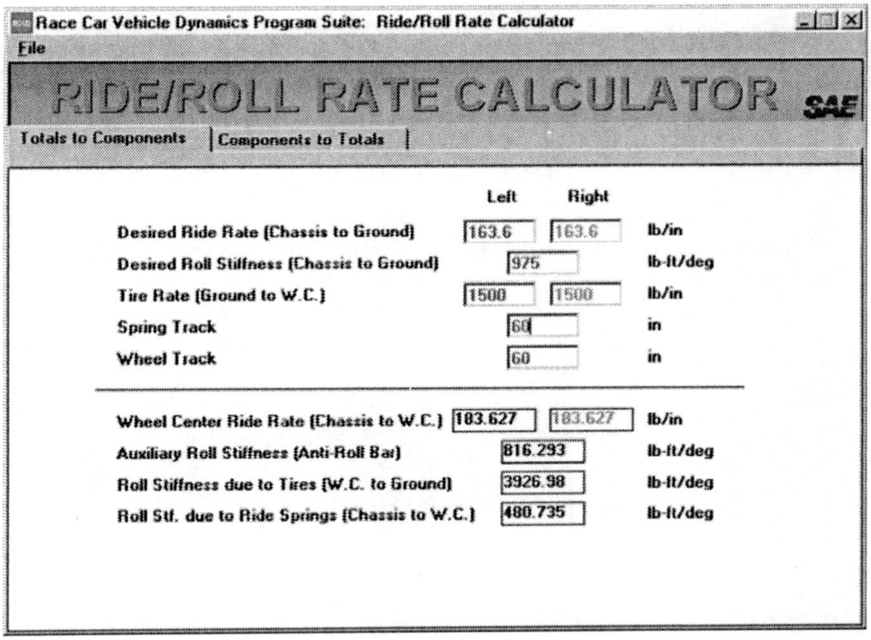

Race Car Vehicle Dynamics Program Suite: Ride/Roll Rate Calculator

File

RIDE/ROLL RATE CALCULATOR SAE

Totals to Components | Components to Totals

	Left	Right	
Desired Ride Rate (Chassis to Ground)	163.6	163.6	lb/in
Desired Roll Stiffness (Chassis to Ground)	975		lb-ft/deg
Tire Rate (Ground to W.C.)	1500	1500	lb/in
Spring Track	60		in
Wheel Track	60		in
Wheel Center Ride Rate (Chassis to W.C.)	183.627	183.627	lb/in
Auxiliary Roll Stiffness (Anti-Roll Bar)	816.293		lb-ft/deg
Roll Stiffness due to Tires (W.C. to Ground)	3926.98		lb-ft/deg
Roll Stf. due to Ride Springs (Chassis to W.C.)	480.735		lb-ft/deg

Figure A.7 *Front springing calculation*

Values are then input into the Ride/Roll Rate Calculator as shown in Figure A.8. A wheel center rate of 134.8 lb./in. and an anti-roll bar providing 109.2 lb.-ft./deg. are needed on the rear.

Note that the spring track on the front is equal to the wheel track since the front has an independent suspension.

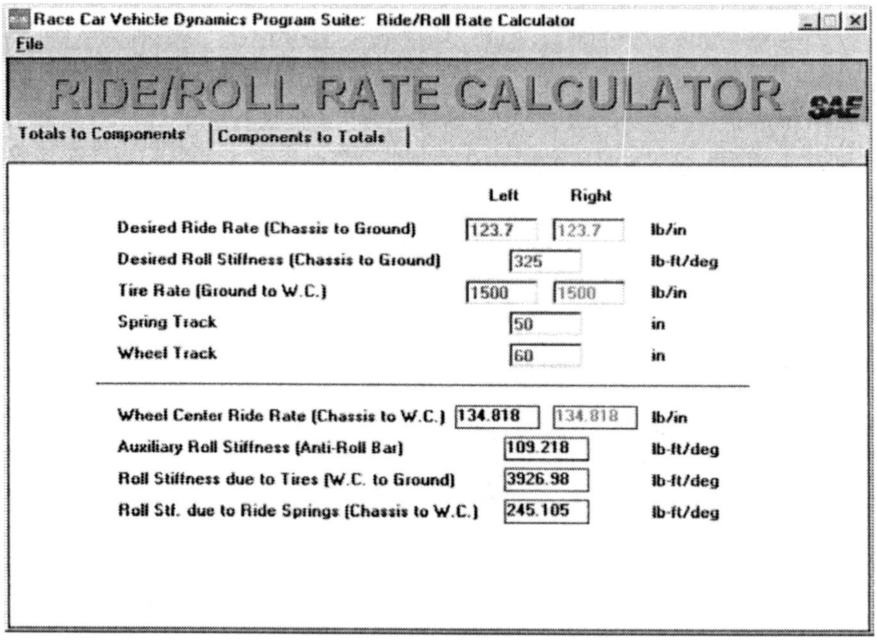

Race Car Vehicle Dynamics Program Suite: Ride/Roll Rate Calculator

File

RIDE/ROLL RATE CALCULATOR SAE

Totals to Components | Components to Totals

	Left	Right	
Desired Ride Rate (Chassis to Ground)	123.7	123.7	lb/in
Desired Roll Stiffness (Chassis to Ground)	325		lb-ft/deg
Tire Rate (Ground to W.C.)	1500	1500	lb/in
Spring Track	50		in
Wheel Track	60		in
Wheel Center Ride Rate (Chassis to W.C.)	134.818	134.818	lb/in
Auxiliary Roll Stiffness (Anti-Roll Bar)	109.218		lb-ft/deg
Roll Stiffness due to Tires (W.C. to Ground)	3926.98		lb-ft/deg
Roll Stf. due to Ride Springs (Chassis to W.C.)	245.105		lb-ft/deg

Figure A.8 *Rear springing calculation*

APPENDIX **B**

Supplemental Material and Bibliographies

B.1 Intelligent Use of *Race Car Vehicle Dynamics*

Race Car Vehicle Dynamics was written for a wide range of readers: college instructors teaching vehicle dynamics, graduate engineers interested in self-study, undergraduate and graduate students, race team engineers and individuals who are intellectually curious about racing cars and vehicle dynamics.

It is useful for the reader to understand the background needed to get the most out of the book before beginning to read it. Initially, significant reviews of necessary background materials were considered by the authors, either for direct inclusion into RCVD itself, or for inclusion in this Problem and Answer book. Upon reflection, we decided that (a) we could never hope to cover the spectrum of materials needed by many of the readers who might be interested in using the book, and (b) other authors had, in fact, already done a good job of doing this in the form of their own textbooks and materials. Consequently, what is presented here is a brief synopsis of the background and prerequisite materials needed to understand the mathematical portions of RCVD and effectively work in the field of vehicle dynamics. Although comprehensiveness was sacrificed, sufficient detail is presented to allow the interested reader to pursue deficiencies, learn new tools and refresh forgotten lessons with a minimum of effort.

The background material needed to study and work in vehicle dynamics consists primarily of four fundamental undergraduate engineering subject areas:

1. Mathematics: calculus, differential equations, linear algebra
2. Dynamics: system modeling, state space, coordinate transforms
3. Controls: transfer functions, feedback controllers, stability criteria
4. Vibrations: free and forced response, eigenvalues, normal modes

In addition, familiarity with the mechanics of solids (including stress, strain and failure modes) and the kinematics of linkages are suggested for the vehicle dynamicist. No mention of computer programming or computer science has been included in the above list. The reason is that it is now an integral part of any engineering curriculum. Engineering software dealing with vehicle dynamics specifically or mathematical solutions in general is abundant—a list of such software and a short description of each package are given later in this appendix.

Any engineering task can be approached either experimentally or theoretically. Vehicle analysis and design typically employs a mixture of the two methods. The vehicle dynamicist usually attempts to theoretically model the fundamental, linear behavior of the vehicle as completely as possible and produces a pencil/paper/computer version of the expected characteristics and performance of the vehicle. Once the fundamentals of a particular proposed vehicle are understood through theoretical modeling, experimental work begins. The goal of this work is to understand unmodeled phenomena, nonlinear behavior and limit performance. So, we can say that we proceed first with theory, followed by experimental verification and extension of theoretical results. We can imagine dynamics and vibrations as fundamental building blocks that are used to develop equations and models of vehicles, describing the vehicle behavior in sufficient detail for the theoretical analyses of interest. We can further imagine mathematics as the tool used for solution of such equations, and control as the tool used for understanding, characterizing and comparing models and solutions. Once the analytical work is complete, experimental work can begin in earnest, though some experimental work is often carried along in parallel with the theoretical analyses.

The typical undergraduate mechanical or electrical engineering student will have studied sufficient mathematics (through differential equations) and will probably have had at least one class in dynamics, control and possibly vibrations. Often a smattering of vibration theory will be included in the controls class in the form of example problems. Much of this material may have been forgotten if not used often. Also, too, many undergraduates leave a controls class with a feeling of abstractness: they can do all of the problems, but they don't understand how control could actually be applied to real engineering problems or issues. There can be a significant conceptual leap required in order to use the lessons of control in vehicle dynamics. The reader should be prepared for this at the outset.

Readers who did not study engineering in college may have never seen many of the topics listed below. In this case, the problem of background becomes a little more interesting. Obviously, the term "review" doesn't apply if the material has never been seen before. Such readers should consider the possibility of either self-study (if extremely motivated) or a return to formal education. Many junior colleges and technical institutes can provide fine courses in the areas of mathematics through differential equations, physics and Newtonian dynamics. However, the subjects of control and vibrations are typically upper-division junior- or senior-year classes which aren't available at such schools and often can't be taken at four-year institutions without enrolling as a full-time student. In such circumstances, self-study is the only road available.

Regardless of where an individual reader falls in the knowledge continuum, the authors recommend perusing the topical list in the following sections as a start. Any review or additional education necessary is probably best undertaken on a need-to-know basis, as the reader progresses through the book. It isn't necessary to have perfect recall, or to learn something about every single item on the topical list. In many cases, only a little familiarity with a concept is sufficient to understand a passage, read a research paper, study a reference or hold a meaningful conversation with a professional colleague. For those who are especially rusty, or are trying to teach themselves the materials, the authors especially recommend the Schaum's Outline Series of books[1] which are not only equipped with significant theory and development but also contain large numbers of solved problems and examples.

B.2 Specific Topics Helpful to the Study of Vehicle Dynamics

Below, we present a list of topics subdivided into five areas: mathematics, dynamics, control, vibration and mechanisms/structures. Some reference books are also given and study/review can be accomplished by referring to these references or other similar books. In the section on mathematics, topics are given only in differential equations and it is assumed that all prerequisite courses in calculus have already been completed. Mathematics through differential equations is serial in nature—each level must be satisfactorily completed before moving on to the next one.

[1]The McGraw-Hill Companies, Inc., http://www.mcgraw-hill.com

B.3 Mathematics

Mathematics deals with the relationships among variables and allows the vehicle dynamicist to formulate solutions, both analytical and numerical, to the equations of motion which describe vehicle motion. In this way, vehicle designs and design modifications can be analyzed on paper before any hardware is constructed, saving both time and money. Because the equations of motion for dynamic systems are differential equations, mathematics training through differential equations is necessary for full understanding of transient vehicle dynamics models and behavior.

1. **Linearity and Superposition** allow results of individual solutions to be added together to produce composite solutions for responses which result when multiple, or complex, inputs are supplied to a system.

2. **Linear Differential Equations** are differential equations in which each term contains a single instance of the dependent variable(s) or one of its derivatives, raised to the power of one. Any dependent variable or derivative raised to any power other than one results in a nonlinear differential equation.

3. **Homogeneous Solutions** to differential equations result when no input is supplied, characterized by the right side of the differential equation equaling zero. This part of the solution is sometimes called the free response.

4. **Nonhomogeneous Solutions** to differential equations result when an input or forcing function is supplied, whether or not the starting or initial conditions are zero. This part of the solution is sometimes called the forced response.

5. **First-Order Differential Equations** are equations in which the highest derivative is one. Linear first-order differential equations/systems can be characterized by a single number, the time constant.

6. **Second-Order Differential Equations** are equations in which the highest derivative is two. Linear second-order differential equations/systems can be characterized by two numbers, the damping ratio ζ and the natural frequency ω_n.

7. **Laplace transforms** are a generalized solution method applicable to linear differential equations with constant coefficients. They transform a differential equation into an algebraic equation. The general methodology for using Laplace transforms consists of three steps:

- Calculate the Laplace transform of the entire differential equation or set of differential equations. This is done by either looking up Laplace transforms in tables or using software such as MATLAB or Mathematica.

- Algebraically solving for the Laplace transform of the dependent variable. Any algebraic methodology can be employed (for example, one could use Cramer's rule, back substitution, matrix algebra, etc.) since the equations are algebraic.

- Calculate the inverse Laplace transform of the dependent variable. This is done using Laplace transform tables or software such as MATLAB or Mathematica. Acquisition of a suitable form for use with Laplace transform tables may require manipulation of algebraic quantities using, e.g., partial fraction expansions.

8. **Fourier transforms** are a generalized method useful for examining differential equations which have inputs with harmonic and/or periodic properties. In general they use mathematical manipulations similar to those employed in use of Laplace transforms: transformation, algebraic manipulation and inversion.

References for Mathematics

1. Bronson, R., *Differential Equations (2nd Ed.)*, Schaum's Outline Series, New York, NY, ISBN 0-07-008019-4, 1994.

2. Edwards, J., *Differential Equations and Boundary Value Problems*, Prentice-Hall Book Co., Englewood Cliffs, NJ, ISBN 0-13-382097-7, 1996.

3. Rainville, E. D., *Elementary Differential Equations (7th Ed.)*, Macmillan Book Co., New York, NY, ISBN 0-02-397-860-9, 1989.

B.4 Dynamics

Dynamics deals with modeling and describing the motion of particles and rigid bodies which have both mass and inertia. We can divide the modeling and development of dynamic equations of motion into two broad, fundamentally different approaches. Traditional or Newtonian dynamics focuses on development of equations of motion through the drawing of free body diagrams. Analytical dynamics develops equations of motion through energy methods. Both methodologies are useful in vehicle dynamics.

1. **Newton's Laws of Motion** describe the general behavior of particles and bodies of finite size and inertia when subjected to external forces and moments. They are the fundamental building blocks used to develop equations of motion.

2. **Free Body Diagrams** are used to separate bodies from one another. Connections are replaced by forces and moments, and Newton's Laws of Motion are then used to develop the equation(s) of motion for the body.

3. **Particles** are bodies possessing mass but no dimensions, hence they have no moments of inertia. Particle dynamics are occasionally used to describe gross vehicle motions, such as in simplified accident reconstruction, momentum balance expressions and the like.

4. **Rigid Bodies** are bodies possessing both mass and nonzero dimensions and thus they have moments of inertia.

5. **Translation and Rotation** are particular kinds of motion in a plane which result from the application of external forces and moments, respectively.

6. **Work and Energy Analysis** is a method for developing equations of motion based on conservation principles rather than on Newton's Laws of Motion.

7. **Impulse and Momentum** is a method for developing expressions for forces and moments in bodies and systems involving variable mass, fluid jets and other non-rigid concepts. It is seldom used in vehicle dynamics analyses.

8. **Periodic Motion** occurs in systems that experience vibrational motion. Motion that repeats itself is called periodic.

9. **Harmonic Motion** is a special class of periodic motion for which the motion not only repeats itself, but repeats itself sinusoidally.

10. **Mass and Moment of Inertia** are properties of a body that cause it to resist translational and rotational motion, respectively. Mass is a function of the total amount of material that the body possesses. Moment of inertia is a function of not only total material but also the placement of that material with respect to a particular axis of rotation.

References for Dynamics

1. Greenwood, D. T., *Principles of Dynamics*, Prentice-Hall Book Co., New York, NY, LCCN 64-19700, 1965.

2. McLean, W. G. & Nelson, E. W., *Engineering Mechanics: Statics and Dynamics (4th Ed.)*, Schaum's Outline Series, New York, NY, ISBN 0-07-044822-1, 1988.

3. Meriam, J. L. & Kraige, L. G., *Engineering Mechanics: Statics and Dynamics (3rd Ed., 2 Vol.)*, John Wiley and Sons, Inc., New York, NY, ISBN 0-471-60295-7, 1993.

B.5 Control

Control deals with analyzing the relationships among variables in a set of equations of motion, often without producing complete solutions for them. Fundamental system properties are studied, such as stability, response time, changes in response to design adjustments and design optimization. System responses to applied and external inputs as well as to actuators driven by feedback controllers are central topics. Control theory provides a number of methodologies that are extremely powerful and universal, are useful in understanding and characterizing vehicle behavior, and for comparing the performance of one vehicle against another.

1. **Input** is the forcing function, command or stimulus applied to a dynamic system. The input may be intentional or unintentional, and may be idealized in a simple fashion (e.g., a step function) or may be very complex.

2. **Output** is the response or output which results from either (a) the application of input(s) or, (b) starting the system from a nonzero or nonequilibrium state. In control, we think of supplying an input and calculating a response; in psychology, we think of supplying a stimulus and measuring a response. There can be multiple inputs and outputs to a complex system.

3. **Transfer Function** is a mathematical relationship between an input and an output. In vehicle dynamics, the input could be intended (steering, braking or a combination of the two) or unintended (wind gusts, road roughness, etc.). The term transfer function implies that Laplace transforms have been used on the equations of motion.

4. **Stability** implies the presence of a restoring force, torque, current, etc. To state that a system is stable means that, if slightly perturbed from its equilibrium position, it will return to that position in finite time, i.e., it has a decaying free response.

5. **Characteristic Equation** is the term applied to the denominator of a transfer function. The name is carefully chosen: the factors or roots of the character-

istic equation determine the character of the system's free response (exponentially decaying, exponentially growing, sinusoid, decaying sinusoid, etc.).

6. **Transients** are responses (to stimuli or inputs) that last only for a finite period of time. If a system is stable, energy will always be extracted over time, so all transients will die-out as time continues unless new input energy is supplied.

7. **Steady-State** is that state reached by a system after all of the transient response has died-out. This could occur slowly or quickly depending upon the system involved.

8. **Steady-State Error** is the difference between the input and output once all transients have died-out. Error is not necessarily bad; a filter, for example, is supposed to produce infinite steady-state error in its rejection frequencies because there isn't supposed to be any output regardless of what the input is.

9. **Tracking Devices** are devices that continually try to make the output equal to the input no matter how the input is changed. Vehicle dynamic handling is a representative tracking problem, in that the driver expects the car to follow his/her inputs faithfully and accurately.

10. **Filtering Devices** are devices that continually try to make the output zero (or at worst a constant value) no matter how the input is changed. Vehicle dynamic suspension analysis is a representative filtering problem, in that the driver expects the suspension system to produce a smooth ride no matter how rough the roadway is. All control systems are either trackers or filters.

11. **Standard Inputs** are inputs or forcing functions that are (a) mathematically simple, and (b) used universally to permit one-to-one comparisons to be made between systems. Examples include step functions, ramp functions, sinusoid inputs, etc.

12. **Root Locus** is a graphical design technique used to determine how the roots of the characteristic equation change as a single parameter is varied. It allows for selection of a value for the parameter which places factors in specified locations and shows how changes in the parameter may cause different roots, transient response, instability, etc.

13. **Frequency Response** is the response of a system when the input is specified to be sinusoidal. All dynamic systems have a frequency response curve and the response of the system to many other types of inputs can be estimated from its frequency response data. Design can also be performed using frequency response data. If, as has been standardized, the output/input ratio expressed in decibels and phase angle expressed in degrees are plotted against $\log_{10}\omega$, the plot is known as a Bode plot.

14. **Design Specifications** are values or system parameters chosen to elicit certain performance characteristics. Specifications are usually given in either the time domain or the frequency domain. Some examples are:

    - *Time Domain Specifications*: rise time, settling time, percent overshoot, mean-square error, etc. (all usually assuming that the input is a step function).
    - *Frequency Domain Specifications*: bandwidth, gain margin, phase margin, etc. (all usually measured directly from a Bode plot).

References for Control

1. DiStefano, J. J. III, Stubberud, A. R. & Williams, I. J., *Feedback and Control Systems (2nd Ed.)*, Schaum's Outline Series, New York, NY, ISBN 0-07-017052-5, 1995.

2. D'Azzo, J. J. & Houpis, C. H., *Linear Control System Analysis and Design: Conventional and Modern (2nd Ed.)*, McGraw-Hill Book Co., New York, NY, ISBN 0-07-016183-6, 1981.

3. Phillips, C. L. & Harbor, R. D., *Feedback Control Systems*, Prentice-Hall Book Co., Englewood Cliffs, NJ, ISBN 0-13-313917-4, 1988.

B.6 Vibrations

Vibrations deals with the periodic, oscillatory motion of rigid bodies and sometimes with the motion of flexible bodies. In many ways, control and vibrations share concepts. Vibrations can be considered to be a special application of control theory in that all of the physical hardware falls into a particular class of mechanical systems. Vibrations can also be considered to be a special class of dynamics. Vibrations can be used to model discrete systems comprised of springs, masses and dampers (most common in vehicle dynamics), or it can produce continuous models (elastic beams, etc.) which are sometimes useful for modeling chassis behavior and other distributed phenomena.

1. **Periodic Motion** is motion that repeats or replicates itself over and over after some fixed amount of time known as the period, T.

2. **Harmonic Motion** is periodic motion that is sinusoidal in nature. Harmonic motion is a special case of periodic motion.

3. **Natural Frequency** is the frequency at which a system oscillates if started from nonzero initial conditions and left alone. The natural frequencies of a system are independent of the inputs supplied, although supplying harmonic inputs at frequencies other than the natural frequency(s) can cause the input frequencies to appear in the output in addition to the natural frequency(s). The natural frequency in rad./sec. is $\omega_n = 2\pi/T$.

4. **Single Degree of Freedom** systems are systems that have only a single mass in motion and possess only a single natural frequency.

5. **Multi-Degree of Freedom** systems are systems with more than one mass and natural frequency. Any system with more than a single degree of freedom is a multi-degree of freedom system.

6. **Stiffness** is that property of a vibrating system that stores energy and later releases the energy back into the system. In many mechanical systems, springs are the stiffness element. Any device which behaves in a spring-like fashion can provide stiffness. For example, tires have vertical stiffness.

7. **Damping** is that property of a vibrating system that extracts or dissipates energy and removes it from the system. In a vehicle, shock absorbers convert kinetic energy of motion into heat, which is rejected to the atmosphere and removed from the suspension system.

8. **Eigenvalues** or characteristic values are the natural frequency(s) of a vibrational system. The term is derived from the German root word "eigen" meaning unique or characteristic. Mathematical methods exist for determining the eigenvalues of a dynamic system, provided the system has been structured into the correct analytical format.

References for Vibration

1. Meirovich, L., *Elementary Vibration Analysis*, McGraw-Hill Book Co., New York, NY, ISBN 0-07-041342-8, 1995.

2. Seto, W. W., *Mechanical Vibrations, Schaum's Outline Series*, New York, NY, ISBN 07-056327-6, 1964.

3. Thompson, W. T., *Theory of Vibration With Applications (2nd Ed.)*, Prentice-Hall Book Co., Englewood Cliffs, NJ, ISBN 0-13-914523-0, 1981.

B.7 Mechanisms and Structures

While RCVD (and vehicle dynamics in general) focuses mostly on gross vehicle motions, lumped-parameter analysis, suspension geometry, etc., the vehicle dynamicist should have a basic understanding of structural mechanics.

1. **Stress** is a force per unit area within a material. Stresses can be normal (tension/compression) or shear.

2. **Strain** is the local deflection, compression or elongation of a material in the presence of stress. It is nondimensional, i.e., deflection (length) divided by the length of the sample.

3. **Young's Modulus** (or Modulus of Elasticity) is the proportionality constant between stress and strain for a given material while it is experiencing elastic deformation.

4. **Elastic Deformation** is an amount of deformation from which the material will return to its undeformed shape once the applied load is removed.

5. **Plastic Deformation** is an amount of deformation from which the material will not return to its undeformed shape once the applied load is removed. The boundary between elastic and plastic deformation is associated with the Yield Stress.

6. **Yield Stress** is the stress at which plastic deformation begins to occur.

7. **Ultimate Stress** is maximum stress a material can sustain before failure occurs.

8. **Fatigue** is the reduction in the ultimate stress of a material due to repeated cycling of loads.

9. **Four-Bar Linkages** consist of three moving links (an input link, coupler link and output link) and a fourth, fixed link usually referred to as "ground". Analysis and design concepts applicable to four-bar linkages are useful in suspension design.

10. **Instant Centers** are the points about which two separate bodies instantaneously rotate. This is a useful concept for velocity analysis of linkages and, in vehicle dynamics, leads to the roll center concept.

References for Mechanisms and Structures

1. Shigley, J. E. & Mischke, C. R., *Mechanical Engineering Design*, McGraw-Hill, Inc., New York, NY, ISBN 0-07-056899-5, 1963.

2. Shames, I. H., *Introduction to Solid Mechanics*, Prentice-Hall Book Co., Englewood Cliffs, NJ, ISBN 0-13-497546-4, 1975.

3. Erdman, A. G. & Sandor, G. N., *Mechanism Design, Analysis and Synthesis*, Prentice-Hall Book Co., Englewood Cliffs, NJ, ISBN 0-13-569872-3, 1984.

B.8 Reference Textbooks in Vehicle Dynamics

The following books constitute a very basic library for the vehicle dynamicist. Each is listed, then followed by a short, subjective evaluation developed by the authors. *All* of these books are worth purchasing and having available as general reference material. Many specific items of interest can be found in them.

1. Wong, J. Y., *Theory of Ground Vehicles*, John Wiley & Sons, New York, NY, ISBN 0-471-03470-3, 1978.

   A general-purpose overview book containing tire dynamics, terramechanics, performance and ride dynamics, steady-state cornering behavior and even some information about tracked vehicles and air cushion vehicles. A good, fundamental reference, but lacking in any transient analyses, particularly in handling dynamics. Now somewhat dated, so specific data for tires, etc., are not applicable to modern vehicles.

2. Gillespie, T. D., *Fundamentals of Vehicle Dynamics*, Society of Automotive Engineers, Inc., Warrendale, PA, ISBN 1-56091-199-9, 1992.

   Similar to Wong, but modernized to include more recent current data. Once more the emphasis, particularly in handling studies, is on vehicle statics, not vehicle dynamics. Chapter headings are similar to Wong. There is also a videotaped version of this textbook material available from SAE.

3. Hucho, W. H., *Aerodynamics of Road Vehicles*, Butterworths Publishers, London, UK, ISBN 0-408-01422-9, 1987.

A first-rate compendium of aerodynamics, just as the title implies. Anyone interested in first-order, or perhaps even second-order vehicle aerodynamic effects and properties should begin here. The references are especially effective and valuable. Written for the aeronautical engineer, and very mathematical in nature, it nevertheless presents lots of practical data which can be used right out of the box.

4. Bastow, D., *Car Suspension and Handling, (3rd Ed.)*, Society of Automotive Engineers, Inc., Warrendale, PA, ISBN 1-56091-4041, 1993.

   A highly practical book which describes suspension parameters in minute detail (springs, shocks, etc.) as well as ride and handling effects produced by suspensions. Written by a designer for designers, thus it is very practical with numerous illustrations and worked-out examples, together with lots of suspension data for particular suspensions.

5. Milliken, W. F. & Milliken, D. L., *Chassis Design: Principles and Analysis*, Society of Automotive Engineers, Inc., Warrendale, PA, ISBN 0-7680-0826-3, 2002.

   Based on the technical notes of ride and handling pioneer Maurice Olley with extensive comments and clarifications by the authors. Olley's career covered a major portion of the development of the modern automobile and he excelled in three disciplines—designer, analyst and experimentalist. These notes were written at the end of his career with General Motors and have been reorganized and converted to book form. Many difficult vehicle dynamics problems are covered in detail, yet the calculations can generally be performed by hand methods—potentially very useful for checking more elaborate computer models.

B.9 Useful Computer Programs for Vehicle Dynamics Work

Notes on Computer Programs: For doing dynamic analysis of vehicle motion, various computer programs are useful. A selection of such programs is listed below, together with some comments regarding the utility of the programs. All are commercially available. Some of these programs (as well as others not mentioned) may be useful in solving some of the problems in this book.

1. MATLAB : A general purpose program with widespread use in the engineering community. Strengths include matrix algebra, systems and control work (especially for problems framed in modern, state-space control terminology) and data visualization. Program capabilities encompass the spectrum of engineering calculations via various "toolboxes" for specific applications, e.g., control systems,

optimization, system identification, etc. The companion program Simulink is an excellent tool for the simulation of dynamic systems.

2. MATHEMATICA: A general purpose, high end, symbolic and numerical methodology for accomplishing a wide variety of mathematical tasks, including numerical integration of equations of motion, Laplace, Fourier, etc., transforms and inverses, symbolic algebra and matrix manipulation, and many others. Widely used in the fields of mathematics and physics.

3. Maple: Similar to Mathematica, Maple is the final word in computer-based, symbolic, mathematical computations. When compared with MATLAB, however, it is not particularly well-suited to numerical computations.

4. MSC/ADAMS: Mechanical system simulation software. It enables users to produce virtual prototypes, realistically simulating the full-motion behavior of complex mechanical systems of links and joints on their computers. Modules are available that include pre-programmed linkage simulations for automotive suspension systems. There are advantages and disadvantages to simulating every component in the suspension if the goal is to study overall vehicle dynamics. Advantages include detailed load and kinematic information for use in stress analysis and design. Disadvantages include lengthy time to build and validate vehicle models and long run times (compared to lumped parameter models discussed below), making parameter studies more time-consuming.

   Other similar kinematics modeling packages are also available, often incorporated directly into mid-range and high-end CAD design software.

5. Computer Algebra Systems: Available on many scientific calculators, such as the Texas Instruments TI-89. These allow for the solution of simultaneous equations, differentiation and integration in symbolic form. While lacking the capability of the previously mentioned programs, computer algebra systems are a valuable assistant to the engineer when performing pencil/paper derivations and calculations.

6. Milliken Research Associates, Inc., licenses a number of specialized programs for vehicle dynamics work. These are comprehensive, elaborate models that have been developed over the last 30+ years and correlated against proving ground tests. These models require a significant amount of data on the chassis and the tires. They are intended for use by the automotive industry, suppliers and professional race teams, and are priced accordingly.

   MRA's models are "lumped parameter" in the sense that the vehicle is described by data that would be acquired from a kinematics and compliance rig. For example, ride-steer would be described by several coefficients instead of detailed specification of the suspension geometry. This technique allows models (input data sets) to be built relatively quickly when compared to more detailed approaches such as multi-body models. Another advantage of lumped-parameter models over multi-body simulations is in run time—as a rule lumped-parameter models solve much faster. This quick turnaround gives the analyst the option of running large parameter studies in reasonable amounts of time. On the other hand, detailed load information for finite element analysis and fatigue prediction may require multi-body simulation. Both types of models have their place in the auto industry.

- Vehicle Dynamics for use with Matlab/Simulink, VDMS: A comprehensive, lumped-parameter, time-based vehicle model programmed in MATLAB and Simulink with 15 principal degrees-of-freedom. Fully nonlinear, it includes kinematics, compliances and aerodynamics and accepts a range of tire models. Intended for both general vehicle dynamics simulation needs and specifically for control system and active system design, much of the code is provided open-source, accessible to the user. VDMS can be compiled for hardware-in-the-loop applications.

- MRA Moment Method, MMM: This unique approach to automotive statics is described in RCVD Chapter 8. The comprehensive model is fully nonlinear. MMM gives the trained user a complete overview of the stability and control of a vehicle with modest effort in modeling and interpretation. We feel that this tool is ideal for initial design studies as well as tracking a design through all the changes that affect handling during the design process.

- Lap Time Simulation, LTS: Simulation to determine the performance of a vehicle on a race course. The comprehensive vehicle model is shared with MMM. LTS is used by race engineers and analysts for "what-if" studies. Modern race cars, and car set-ups, are very carefully optimized. LTS and similar lap simulation programs have been developed to supplement track testing time with simulation. Once the LTS model (data set) has been validated for a particular race car, it allows many different set-up combinations to be tested at very low cost when compared to track testing.

- Vehicle Dynamics Simulation, VDS: Similar to VDMS, but not programmed in the MATLAB environment, VDS is a comprehensive, lumped-parameter model with its origins in the work at CAL in the late 1960s. Includes terrain map, a driver model and a detailed steering system model.

- MRA is regularly asked to model unique vehicle dynamics problems, for example, the addition of active control elements to a vehicle. This can be done either through the modification of one of the comprehensive models (above) or through the production of a small, special purpose model just for the project. Milliken Research Associates, Inc., can be contacted through their website: http://www.millikenresearch.com

7. RCVD Program Suite: The accompanying CD includes a suite of elementary vehicle models. These are useful as an introduction to vehicle simulations and as a vehicle dynamics learning tool. While limited in scope and capability, they demonstrate many of the capabilities of more elaborate models.

B.10 Sources for Vehicle and Tire Data

Data for vehicle dynamics modeling is available from several sources.

1. There are a number of papers that summarize vehicle parameters, some of these are:

    a. Rasmussen, R. E., F. W. Hill, P. M. Riede., "Typical Vehicle Parameters for Dynamics Studies", General Motors Publication A-2542 (or A-2541), General Motors Proving Ground, Milford, Michigan, USA, April 1970.

    b. Basso, G. L. "Functional Derivation of Vehicle Parameters for Dynamic Studies", Laboratory Technical Report LTR-ST, 747, National Research Council Canada, September 1974.

    c. Riede, P. M., R. L. Leffert, W. A. Cobb., "Typical Vehicle Parameters for Dynamic Studies Revised for the 1980s", SAE Paper No. 840561, Society of Automotive Engineers, Warrendale, PA, 1984.

    d. Rompe, K., E. Donges, "Variation Ranges for the Handling Characteristics of Today's Passenger Cars", SAE Paper No. 841187, Society of Automotive Engineers, Warrendale, PA, 1984.

    e. Barak, P., "Magic Numbers in Design of Suspensions for Passenger Cars", SAE Paper No. 911921, Society of Automotive Engineers, Warrendale, PA, 1991.

    f. Garrott, W. R., "Measured Vehicle Inertial Parameters — NHTSA's Data Through September 1992", SAE Paper No. 930897, Society of Automotive Engineers, Warrendale, PA, 1993.

2. Manufacturers Motor Vehicle Specifications – these are prepared and distributed by vehicle manufacturers. They are available for each car model and follow a stan-

dard format provided by the manufacturers' trade association. They are available from the individual car companies, both US and worldwide.

3. A source for more recent vehicle data is the Internet. We believe that some of the major automotive companies are now posting their Manufacturers Motor Vehicle Specifications on their corporate World Wide Web site(s). We hope that this will become a universal practice in the future.

4. Tire data is extremely difficult to find for a variety of reasons. While tires may appear quite simple on the outside, they are extremely complex to design and manufacture. Historically, tire manufacturers have tended to protect many of their developments as trade secrets instead of patents. Tire data of all forms beyond basic size information must be developed by a variety of expensive testing procedures and (with the exception of load range and other legislated data) is usually not published. Due to the large expense of testing, detailed data is available on relatively few of the huge variety of tires available in the market. Limited tire force and moment data has been made available by the various tire companies. For example, Goodyear has published some data for their Formula SAE tires on the web.

B.11 Sample Course Outlines

Race Car Vehicle Dynamics is being used as a text in several university courses. Two course outlines are given below.

University at Buffalo, State University of New York

The "Road Vehicle Dynamics" course at the University at Buffalo (UB), taught originally by Prof. Bill Rae and now by Edward Kasprzak, was first offered in 1996. This one-semester, three credit-hour course is intended for seniors and graduate students. The class meets twice weekly for one hour and thirty minutes, with a total of 28 meetings throughout the semester.

The following syllabus, originally developed with the assistance of the Millikens, has been refined over the years. It accounts for 24 lectures, leaving sufficient time for two exams, reviews and a guest lecturer [2].

- **Introduction—Chapter 1**: One lecture. Combined with the usual first day material (explanation of course policies, syllabus, grading procedures, etc.), this chapter makes a good introductory lecture. Focus on accelerations and the "g-g" diagram.

[2] A more detailed syllabus is available at www.millikenresearch.com/rvd.html

- **Axis Systems—Chapter 4**: One lecture. Allows for the introduction of a large amount of vehicle dynamics terminology which will appear throughout the course.
- **Tires—Chapters 2 & 14**: Five lectures. Generation of lateral force and aligning torque as a function of slip angle and inclination angle. Induced drag. Longitudinal forces and the slip ratio. The friction ellipse. Nondimensional approach to tire data. Tires are the principal force/moment generators for vehicle maneuvering, so a thorough investigation of these chapters is essential.
- **Steady-State Stability and Control—Chapter 5**: Five lectures. Slip angle generation by the vehicle, equations of motion, stability derivatives, over/understeer. Incorporate the RCVD Program Suite (Bicycle Model and RCVD Speedway) at this point to supplement RCVD. Steady-state responses, significant speeds. This chapter is central to the course.
- **Transients—Chapter 6**: One lecture. This chapter is mostly a review of material students will have already encountered at UB, so a single lecture relating their knowledge of transient response to the automobile is sufficient.
- **Four-Wheeled Vehicle Models—Chapters 7, 16, 18**: Five lectures. Calculation of static wheel loads, load transfers, roll stiffness, pair analysis, ride, etc. Use of the RCVD Program Suite (Four-Wheeler, RCVD Speedway 2, Ride and Roll Rate Calculator) is very useful here. By the end of these chapters the student should have a good handle on the dynamics of real automobiles.
- **Force-Moment Analysis—Chapter 8**: Two lectures. The notion of quasi-static analysis of a dynamic system is foreign to most students (except those with a good understanding of aircraft stability and control analysis). This chapter is very useful for both the overall concept and the wealth of information which Force-Moment provides about the vehicle.
- **Race Car Design, Set-up and Tuning—Chapters 9-12**: Two lectures. Here is a chance for students to "put it all together" through the investigation of how changes in vehicle parameters affect handling. While the focus can be on race cars, the instructor should be mindful to keep passenger car performance (linear range) in mind. Students generally enjoy the very practical nature of these lectures as opposed to the "engineering theory" typical of most engineering courses.
- **Other Topics**: Two lectures, interspersed throughout the semester, are spent on other topics such as aerodynamics, steering systems, suspensions, impact (crash) dynamics, drivetrain, etc. Suspension and steering design merits a course of its own, and the book *Chassis Design: Principles and Analysis*[3] surpasses RCVD as a text for detailed investigation of these topics. At the time of writing, UB was considering adding a second vehicle dynamics course to cover these topics.

[3] *Chassis Design: Principles and Analysis* by William and Douglas Milliken, SAE, R-206, 2002.

University of Illinois, Urbana-Champaign

Prof. L. Daniel Metz provides the following course outline, based on his extensive teaching experience at University of Illinois, Champaign-Urbana and SAE Seminars.

The "standard" semester course is three credit hours and meets for 50-minute periods three times/week for 15 weeks, a total of 45 lectures. Subtract two or three lectures for examinations and you are left with about 42 actual lecture periods in which to cover materials. Taking out the equivalent of perhaps four lectures to cover homework problem solutions and work out some sample problems in class leaves 35 net hours to convey materials. Bear in mind that a course in vehicle dynamics would be an elective course and that students are reluctant to take electives which are less than three credit hours/semester.

- Tires, Chapter 2 – four lectures covering the entire chapter. I would neither skip nor minimize anything in Chapter 2; I think that tire mechanics is the bedrock on which vehicle dynamics is built and I would take my time and cover that chapter extremely thoroughly.

- Vehicle Axis Systems, Chapter 4 – four lectures covering the chapter plus the topics listed in the four sections of Supplementary Materials on Mathematics, Dynamics, Control and Vibration. My experience with undergraduates and graduate students is that even though they may have taken (or been taken by!) the prerequisites for a course, they don't remember much, particularly when it comes to control and vibration. I always have to review these subjects when I teach vehicle dynamics here or for SAE. For these four lectures, we wouldn't actually be using much from RCVD, but they set the tone for students and make them realize that the prerequisites are for real and that they will need them. If review by a student is necessary, (s)he is forewarned here.

- Simplified Steady-State Stability and Control, Chapter 5 – seven lectures covering the entire chapter. For most students, this will be a significant part of what they need and expect to get out of the course. I think every topic in that chapter is valuable and if time is short, this isn't the place to skimp. Chapters 5 and 6 will form the building blocks for any student who actually wants to practice vehicle dynamics in his/her profession.

- Simplified Transient Stability and Control, Chapter 6 – six lectures covering the entire chapter. All the comments regarding Chapter 5 above apply here as well.

- Steady-State Pair Analysis, Chapter 7 – two lectures covering the entire chapter. This is a relatively simple concept for students to grasp once they have covered everything above it and it leads naturally to limit performance steady-state analysis via the MMM. I think it is important enough to do something with, albeit not too much.

- "g-g" Diagram, Chapter 9 – two lectures covering the entire chapter. Another very fundamental chapter for the analysis of real driving maneuvers, that is, maneuvers that involve steering, braking and acceleration simultaneously. Too important to leave out. The "g-g" diagram is one of the fundamentals of understanding actual (as opposed to highly idealized) experimental engineering of vehicles.

- Suspension Geometry, Chapter 17 – four lectures covering the entire chapter. I like for students to study suspension geometry and see it as a way of using kinematics to adjust tire attitudes, with the associated changes in stability and control which result (either intentionally or accidentally) from those adjustments.

- Steering Systems, Chapter 19 – two lectures covering the entire chapter. I'd cover this chapter quickly, but the segments on roll steer and alignment shouldn't be overlooked.

- Ride and Roll Rates, Chapter 16 – four lectures covering the entire chapter. Too often vehicle dynamics is taken to be a synonym for handling dynamics. In this chapter, some attention can be paid to the question of ride. The compromises between handling and ride necessitated by a passive suspension system can be emphasized and vibration theory can come to the fore.

  The above outline is 35 lectures, if more time is available my (Dr. Metz's) next choices would be:

- Wheel Loads, Chapter 18 – four lectures covering the entire chapter.

- Force-Moment Analysis, Chapter 8 – six lectures covering the entire chapter. I've left the MRA Moment Method (MMM) until last, but not because I don't think it's important. Far from it! But I do think it is a complex methodology which can't be usefully employed without a solid background in the fundamental issues. So the order chosen is intended to provide that background. If time permits, it would certainly be on my list.